Table of Contents

Tipple at Columbine Mine, Erie
Rocky Mountain Fuel Co. operated Columbine Mine from 1920 until it was closed in 1946. The mine was located south of Erie and the mine camp of Serene was close by for housing for the miners. 7,316,275 tons of coal were produced at the mine.
Photo from Louisville Public Library and Louisville Historical Museum

Coal Mining *in* Colorado's Northern Field

Carolyn Conarroe

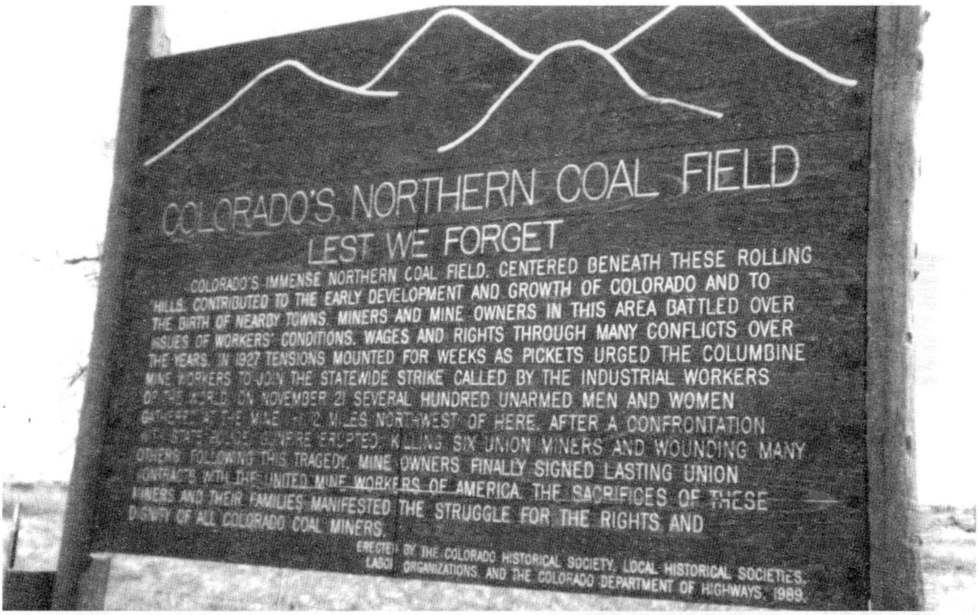

(Top) Sign erected by the Colorado Historical Society, local historical societies, labor organizations, and the Colorado Department of Highways in 1989 and is four miles east of Lafayette on Highway 7. *(Below)* An area of Colorado's Northern Coal Field as it appears today.

Published by Conarroe Companies
Designed by Michael Campbell
Printing by Centennial Printing

First Edition
Printed in the United States of America

ISBN 0-9711073-1-9

Library of Congress Control Number: 2001090796

Introduction

The settlers found the coal along the creek bank and warmed their homes during their first winters in the territory. They named the creek Coal Creek. Farther east along the basin of Coal Creek the homesteader plowed his fields and turned up coal.

The coal was there; the mining followed. However, the opening of the coal deposits in the Northern Coal Field began slowly. Not until the railroads came through the country north of Denver could the product be moved economically to a market. When the railroads were in place, the mines opened for full production and the miners arrived. Coal mines from Marshall to Frederick were in production from 1856 until 1975.

The Northern Coal Field created a work opportunity and a way of life that holds its own place in the history of Colorado.

My interest in the history of Louisville, Colorado, and the coal mining heritage began in 1965 when our family moved to the area.

Who could know that coal mines lay under the rural landscape of east Boulder County. Mine tipples and buildings were gone and features erased. The red ash heaps looked like mounds of dirt.

Slowly friends in Louisville began to talk about the history. I had the good fortune to be co-publisher of Louisville Times and also to be working in the Louisville Public Library during the 1970's. And at that time the old timers were willing to talk to me about their histories.

Research grew and photographs were collected and by publishing the information in columns in the Louisville Times, the story of Louisville was put together and errors corrected. In 1978 "The Louisville Story" was published.

It soon became apparent that the other coal mining communities, including Lafayette, Superior, Erie, Frederick, Firestone and Dacono, had undergone the same mining experiences and their histories are intertwined.

The towns grew and prospered in relation to the coal mining activities and market. They were essentially democratic communities, not dominated by mine owners and companies. They struggled alike in the campaigns for union recognition and through the strikes and strife. When the coal markets were gone, the towns removed the evidence of the coal mining history and waited.

With the surge in growth along Colorado's front range, the coal mining towns — their heritage hidden — suddenly became suburban communities with new families, new houses, and modern conveniences that those people living here in 1880 could not have imagined.

Memorial statue to Erie and Northern Field miners, Erie, Colorado.
(inset) Memorial statue to Louisville miners, Louisville, Colorado.

Other books and many articles have been written about the coal mines, the wars and events, and the people involved. A list of some of the books and articles which have been published and are recommended for further reading is in Sources *(pages 74 & 75)*.

Research for 35 years has given me boxes of information which I wanted to organize into this book about the Northern Coal Field. It's easier to buy a file cabinet, but a book provides an opportunity to look again at the history of the mines and towns in this area.

So that new residents will know the history and so that old timers will not be forgotten, this book is offered.

Acme Mine, Louisville

ACME MINE WAS OPENED IN 1890 AT A DEPTH OF 186 FEET. IT WAS CLOSED IN 1928 AFTER PRODUCTION OF 1,729,282 TONS OF COAL. OPERATORS WERE LOUISVILLE COAL MINING CO., UNITED COAL CO., AND NORTHERN COAL & COKE CO., WHICH SOLD MINES IN 1911 TO ROCKY MOUNTAIN FUEL CO. ACME MINE UNDERLIES CITY OF LOUISVILLE AND WAS WORKED ON TWO LEVEL.
Photo from Louisville Public Library and Louisville Historical Museum

Matchless Mine, Louisville

Located west of Louisville, Matchless Mine was operated by American Fuel Co. from 1903 to 1912. Matchless Fuel Co. operated the mine until it was closed in 1927. Worked at 238 feet, it produced 559,228 tons of coal.

Photo from Louisville Public Library and Louisville Historical Museum

Chapter 1

Opening Northern Field

The Northern Field is in the Laramie geologic formation, a portion of the large Denver Basin, and today is known as Boulder-Weld Coal Field. The coal field extends from the area known as Marshall in Boulder County northeasterly toward the town of Frederick in Weld County. The Laramie formation, which underlies an estimated 87,000 acres of land in northern Colorado, includes seven layers of coal of mineable thickness. The coal layers are mixed through claystone, shale and sandstone.

A Colorado Geological Survey report on the characteristics of the coal layers in the Northern Field is summarized :

Coalbeds #1 and #2 are too deep to be mined.

Coalbed #3 is the main, or Gorham seam, and was most extensively mined in the Imperial Mine east of Erie. The Gorham seam ran to 14 feet in thickness.

Coalbed #4 runs one foot to 35 feet above Coal bed #3. In Marshall and Lafayette Districts, Coal-beds #3 and #4 merged into one bed of up to 12 feet and the deposit was mined extensively. The Columbine Mine south of Erie also mined the #3 and #4 seam.

Coalbed #5 was the middle seam and was located generally 35 feet to 8 feet above Coalbed # 3. The mines in the Frederick and Dacono District were in this seam.

Coalbed #6 ran 120 feet above Coalbed #3. This seam was mined near Erie, generally underground, but a coalbed at this level could be strip mined.

Coalbed #7 is the uppermost layer, thin and not compacted because it is so close to the surface. This coal was not mined commercially.

The coal in these seams or veins was a sub-bituminous or lignite, a much softer coal than anthracite, the hardest, or bituminous coals. This lignite-type coal had a high moisture content. When the coal was brought to the surface it began to dry out and if allowed to sit about for long it began to crumble.

Rock faults were found at intervals throughout the coal field and formed natural barriers in the coal mines. Many of the mines were opened for the purpose of being able to work around the faults.

The settlers who first reaped the benefits of the area's coal warmed their houses when they arrived to settle homesteads.

Coal mining in the Northern Field began in 1856 in an area just east of the foothills of the Rocky Mountains. In 1866 coal was discovered 25 miles farther east and the town of Erie was settled. In the 1860's the coal from these mines was transported to the Denver market by team and wagon. The price of coal was high

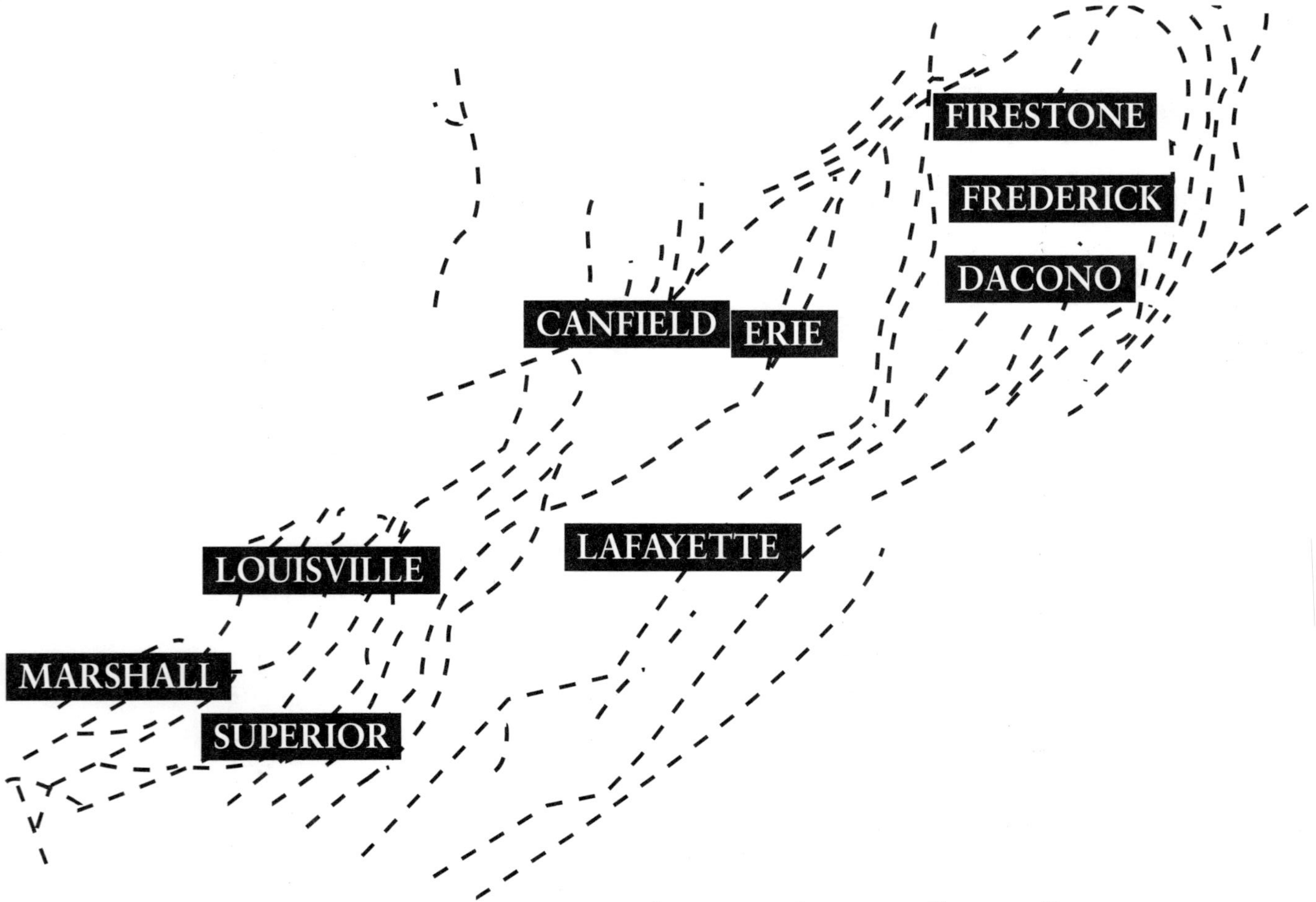

LOCATIONS OF THE ROCK FAULTS IN THE FOUR QUADRANGLES - LOUISVILLE, LAFAYETTE, ERIE AND FREDERICK.
From records of Colorado State Mine Inspector and Colorado Geological Survey

and although the market was there the costs of production and transportation were too high to provide profit.

William A. Kitchen opened the mine he called the Washington Lode on Coal Creek in 1856. It would be two decades before a railroad would arrive to transport the mine's product to market, and Kitchen could not anticipate the importance the coal would have to the economy of the territory. In 1866 Kitchen sold his mine to Joseph Marshall.

The opening of the Northern Field waited for the railroads to come into Boulder and Weld counties.

Railroads

The first railroad to prompt the development of the coal mining industry in the Northern Field was the Denver, Utah & Pacific Railroad. A spur was put down from Brighton into Erie in 1871.

A coal vein had been opened by Ira Austin in 1866 on a hillside east of an area that was to become Erie. With the arrival of the railroad the cluster of homes near the mine was platted in 1871 and incorporated as the Town of Erie in 1874.

At about that same time mines to the west of Erie and the surrounding homes became the Canfield community. In 1873 the Denver, Utah & Pacific Railroad spur was continued through Canfield and on to Boulder, growing from spur to rail line. The line was later acquired by the Denver & Boulder Valley Railroad and eventually taken over by Union Pacific. The Denver, Utah & Pacific line was also expanded to run to Longmont through Canfield.

A narrow gauge route called Longmont's Baby Railroad began operating between Longmont and Erie in 1881 to move coal to market. Colorado Southern took over the right-of-way when the narrow gauge ran into financial difficulties. The narrow gauge track was converted to standard gauge in 1883. In 1889 it was sold to Chicago, Burlington & Quincy Railroad, a company which later became the Burlington Northern Railroad.

The Colorado Central Railroad was organized by a business group from Golden.(1) Boulder County issued $150,000 worth of bonds to finance the railroad into the county. The Colorado Central tracks came from Boulder and crossed Coal Creek. The tracks were put down in 1872 and 1873 and the Coal Creek depot was opened. The town that was founded near the depot was Louisville. C. C. Welch, on the Board of Directors of the railroad, was the first to take advantage of the discovery of the coal the farmer, Dave Kerr, had plowed to the surface. Welch financed the opening of a coal mine on the Kerr property which was to the east of the depot. Kerr received a royalty for the coal taken from his property.

Joseph Marshall chartered the Marshall Coal Mining Co. when the Golden, Boulder and Caribou Railroad opened a line from Boulder to Marshall's mine. The first car load of coal to be shipped over the new rail line was on March 5, 1878. The Denver, Marshall and Boulder Railroad opened another line in 1885 for a direct route to Denver. The line was acquired by Colorado Central and later became Colorado Southern Railroad.(2)

Simpson Mine, Lafayette

The mine was opened by Simpson Brothers in 1887 on land leased from Mary Miller. In 38 years Simpson Mine produced a reported 4, 129,693 tons of coal. The mine closed in 1926.

(Photo from Glenn Lybarger) From collection of Lafayette Miners Museum

With at least three railroad companies convinced of the economic opportunities opening up in the new coal field and rails laid to the areas of mining activity, the Northern Field was open for business. The communities that had been growing up around the mines expanded along with the mining industry and the transportation.

By 1881 the mining communities of Marshall, Louisville and Canfield-Erie were principal centers and had produced 247,000 tons of coal.

The date of record for the opening of the Baker Mine on Coal Creek is 1881. Mary Miller said in a 1912 Lafayette Leader story that coal was discovered in 1873. The Baker Mine was in an area east of the Miller farm, the site of future Lafayette.

The Simpson brothers, John H. and James Jr., closed their store in Louisville and bought farmland a mile to the east. They prospected for coal on Section 2, the farm of Mary Miller, and leased the ground to open the Simpson Mine in 1887. However, there was no serious production from either the Baker Mine or the Simpson Mine until 1888 when a spur of the Colorado Central Railroad connected those mines to the Louisville Depot.

A town was being settled around the mines in 1884 but it was not until 1889 when the Denver, Utah & Pacific Railroad ran tracks from Erie, past the east side of Lafayette that the mines and the town could grow. The railroad continued to Denver but a branch of the railroad came into and through the town as well.(3)

Two mines in Superior were opened by 1895, the Industrial and the Enterprise. The Gorham Slope mine has no date of record for opening. As the Colorado Central Railroad passed through the area from Marshall to Denver, the Superior coal (so named by settler William Hake who was convinced of its superior quality) enjoyed a market from the town's early days.

Frederick, Firestone and Dacono, all in Weld County, joined in the coal mine development when McKissic Mine opened in 1887. Union Pacific Railroad helped to open the coal mines in the eastern most part of the Northern Field.

RAILROAD OWNERSHIP SEQUENCE:

Colorado Central RR > Colorado Southern RR

Colorado Southern RR > Chicago, Burlington & Quincy RR > Burlington Northern RR
> Burlington Northern Santa Fe RR

Denver Utah & Pacific RR - and - Denver & Boulder Valley RR > Union Pacific

Railroad car loaded with coal from Acme Mine, Louisville, Colorado.
Photo from Louisville Public Library and Louisville Historical Museum

Map of Railroads in Northern Field of Colorado

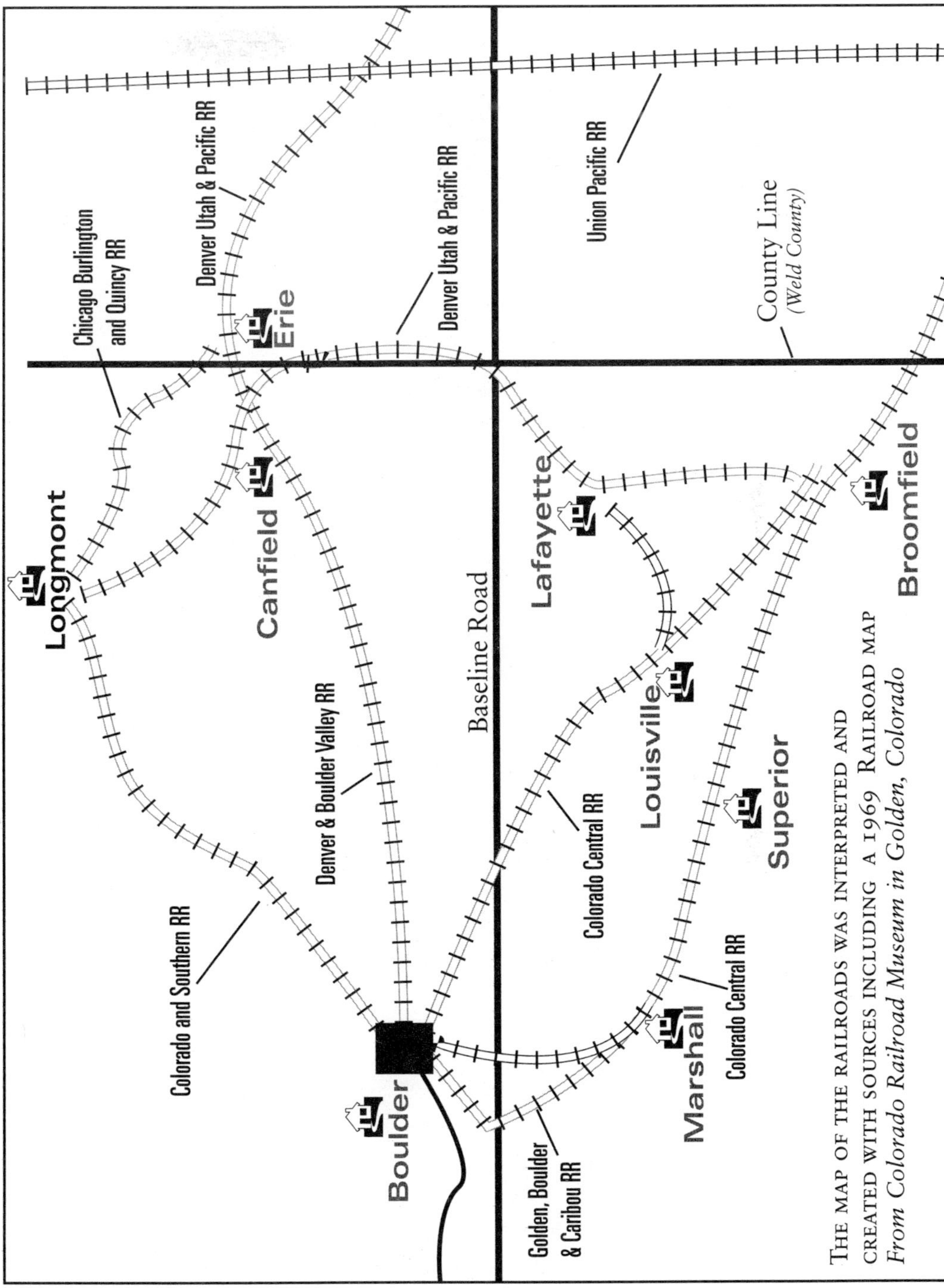

THE MAP OF THE RAILROADS WAS INTERPRETED AND CREATED WITH SOURCES INCLUDING A 1969 RAILROAD MAP *From Colorado Railroad Museum in Golden, Colorado*

Crown Mine, Superior

The Crown Mine at Superior operated from 1938 to 1955. Production was 618,413 tons of coal. The shift photo was taken May 12, 1942.

Photo from collection of the late Leon Coet.

Chapter 2

Towns and Mines of Northern Field

The Marshall area, south of Boulder, Colorado, records the first mine opening, but Erie was the first town in the Northern Field to develop around the coal mines.

The Briggs Mine opened in 1866 near a stagecoach stop. Miners had begun to settle near their mine, and the railroad spur from Brighton came to the mine settlement in 1871. With the industry, workers and their families, and the transportation, Erie grew and was incorporated in 1874. Boulder Valley Coal Company was organized to operate the mine before 1875 and this company alone shipped 175 tons a day. Other mine companies were organized, and 200 men were employed in the busy winter months.

Erie was never a mining camp or company-owned town. Residents provided their own government and independent businesses arrived to provide goods and services. A school had been opened for the territory in 1861, and Erie built its first school in 1874. Churches were founded and grew with the town.

The pattern for independent communities, schools and churches followed in Louisville, Lafayette, and all the towns in the Northern Field.

Canfield mines were opened within a few years after the Erie mines were in operation. The town of Canfield was platted in December 1875 by Charles Wallis, William Wise, B. M. Williams and J. O. Wise. The Jackson Mine, one of the first mines at Canfield, was owned by the Jackson Coal Co. and managed by D. S. Woods. W. O. Wise managed and was part owner of the Star Mine. In 1880 the community reported a population of 200. Businesses came to the town and daily mail delivery was offered in 1882. In spite of the activity and promise, Canfield never did incorporate or set up a town government. [4]

Joseph Marshall's mine was in the largest seam of coal in the Northern Field. Miners were working there in 1866. In 1878, when the railroad could carry the mine's production to market, other mines began to open and businesses moved to the area to provide services. The developing community became known as Marshall but was never incorporated.

Louis Nawatny, who had been the surveyor for the Colorado Central Railroad when it crossed Coal Creek, had stayed in the area to engineer the opening of the area's first mine, the Welch Mine in 1877. Nawatny also bought land west of Coal Creek Depot, surveyed his property and platted a town which he named for himself. Louisville was incorporated in 1878. The Welch mine, later called the Louisville Mine, was the biggest mine in the area by 1886. A fire from spontaneous combustion closed the mine for two months. Marshall consolidated Coal

Company bought out the coal seam and opened Louisville #2.

Mary Miller had begun to anticipate the growth in the area of the Baker Mine and Simpson Mine [5] which had been opened on her land. In 1884 she set aside 150 acres of her farm for a new town. She platted the town in 1885 and named it after her late husband, Lafayette. The plat was filed Jan. 31, 1888, and the town incorporated in 1889. Another mine, the Cannon Mine was opened on the Miller farm in 1888. Mary Miller collected royalties but her interests were in the business of the town and the properties, not in coal mining. [6]

With her parents, John B. and Harriette Foote, she deeded lots for homes all through her plat, and in 1902 for Lafayette and Louisville Electric Co. on the west of her property.[7] On the east side of town she deeded land for the Denver, Utah and Pacific Railroad in 1888, and in 1899 she sold lots, also on the east side of town, to James Kirk for $35,5000. In the deed she stipulated a ban on the sale of beer and wine, actually spiritous, vinous or malt liquors.

One of the first mines to be opened in the Superior area was the Enterprise Mine in 1895. According to one report an earlier mine, the Hood Mine, was opened by James Hood and George Morrison in 1892 on the property of William Hake. The report says the Hood Mine was sold in 1898 to the Northern Coal and Coke Co. and the name was changed to Industrial Mine. The date of record for the opening of the Industrial Mine, which was the largest mine in Superior, is 1895. Hood is on record as sinking the shaft for the Industrial Mine and was an officer of the Industrial Mine Co. Mines of the Northern Coal and Coke Co. were taken over by Rocky Mountain Fuel Co. in 1911.

The town of Superior was platted in 1896 by William Hake and was incorporated in 1904. Businesses opened and homes were built around the mine as well as in the area of the businesses. Superior was one of the few coal camps in Northern Field to have a company store. After the Industrial Mine closed in 1945 many of the houses were moved away. For 50 years Superior was one of the smaller towns in Northern Field. [8]

McKissic Mine opened in 1887 near Firestone. However, the towns of Frederick, Firestone and Dacono in Weld County were settled later than most of the Boulder County towns in the Northern Field.

Dacono incorporated in 1901. C. L. Baum, president of the large Baum Mine, named the town for his wife, Daisy, and her two friends, Cora and Nora. Frederick incorporated in 1907 and Firestone incorporated in 1908. Both towns were named after the owners of the land on which they were founded: Frederick Clark and Jacob Firestone. The coal mines of the tri-towns were later in getting into full production but Lincoln Mine was still mining coal in 1974. Miners from all over the area, and as far away as Louisville, continued to find employment near Dacono.

Over 195 mines were opened in the Northern Field and possibly more which have not been identified.

To see the extent of these 195 mines and understand more the effect the mines had on the surrounding land, the economy and the towns, the locations have been plotted on maps for each community. These six maps have been interpreted and created from information on Colorado Geological Survey maps of mines and coal deposits in the Northern Field.

Industrial Mine, Superior
Industrial mine was worked from 1895 to 1945 and was Superior's largest mine. Rocky Mountain Fuel Company operated the mine from 1911 to its closing.

Rex Mine #1, Louisville
Rocky Mountain Fuel Co. operated Rex #1 at Louisville from 1911 until the mine was closed in 1917. It was opened in 1898 at 158 feet. The mine produced 1,666,217 tons of coal.
Photo from Louisville Public Library and Louisville Historical Museum

Maps of towns and coal mines - *E 1 through 6*

Marshall

Allen Bond Slope	1890 - 1895	Lewis #1	1914 - 1925
Big Five	1938 - 1942	Lewis #2	1934 - 1942
Black Diamond	No date	Marshall #1	1856
Black Diamond (Old)	No date	Marshall #2	1856
Black Diamond #1	No date	Marshall (Old)	1856
Blue Goose	1921 - 1924	Marshall Shaft	1856
Blue Ribbon	1905	Mitchell	1893 - 1897
Cactus	1922 - 1935	Murray Slope	No date
Coal Creek Slope	1935	Pinecliff	1929 - 1942
Cook Shaft	1928	Pittsburg	1926 -1952
Cracker Jack (Old)	1917 - 1948	Peerless	1926-1933
Cuba	1957 - 1958	Premier	1933
Eagle (Old)	1934	Red Ash #1	1916 -1926
El Dorado	1933-1939	Red Ash #2	1916-1926
Fox	1880 - 1936	Ross (Old)	1930-1939
Fox (Old)	No date	Ross (New)	No date
Fox #6	1899	Rosser	1897-1898-1907
Fox Slope	No date	Rosser (New)	1928-1929
Glo Coal	No date	Rosser #7	1899-1906
Gorham	1898 - 1939	Rowley Slope	No date
Highview	1930 - 1936	Shanahan	1897-1907
Imperial (Old)	1895 - 1898	Sunrise	1927-1936
Kitchen Slope	No date	Tropic	1935-1936
York Strip	1940 - 1946		

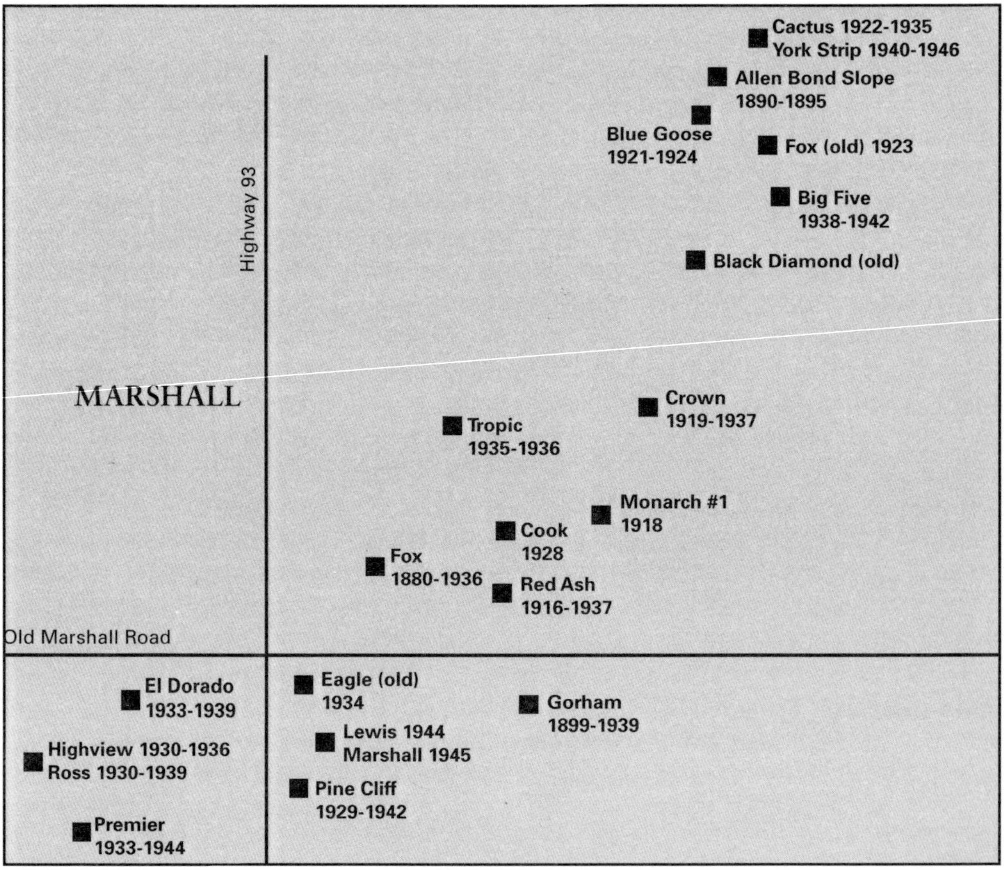

Lousiville

Abadair Link Slope	No date	Liley	1938 - 1948
Acme	1888 - 1928	Lucus	1908
Acme Air Shaft	No date	Marshall Shaft	No date
Ajax	1890 - 1892	Matchless	1903 - 1918
Big Lake (Garibaldi)	1916	Monarch #1	1902 - 1918
Caledonia	1890 - 1899	Monarch #2	1909 - 1947
Centennial (Old)	1906 - 1931	Nonpariel	1907 - 1925
Centennial (New)	1917 - 1921	Northern	1938
Champion	1917 - 1921	Paramount- DiGiacomo	1933 - 1939
Crown (New)	1938 - 1955	Paramount- Domenico	1906 - 1922
Davidson	1888 - 1892	Pennsylvania Shaft	No date
Electric	1898 - 1918	Pluto #1	1896 - 1933
Excelsior	1890 - 1913	Regal	1935 - 1947
Eversman	1928 - 1929	Rex #1	1898 - 1917
Fireside	1931 - 1944	Rex #2	1898 - 1915
Hartman	1927	Summit	1909 - 1913
Hecla #1	1890 - 1920	Sunland	1909 - 1913
Hi Way	1930 - 1954	Sunnyside	1900 - 1921
Link Slope	No date	Vulcan	1903 - 1937
Imperial	1895 - 1898	Welch - Louisville	1877 - 1891
Leader	1893 - 1899	Wilson Shaft	No date

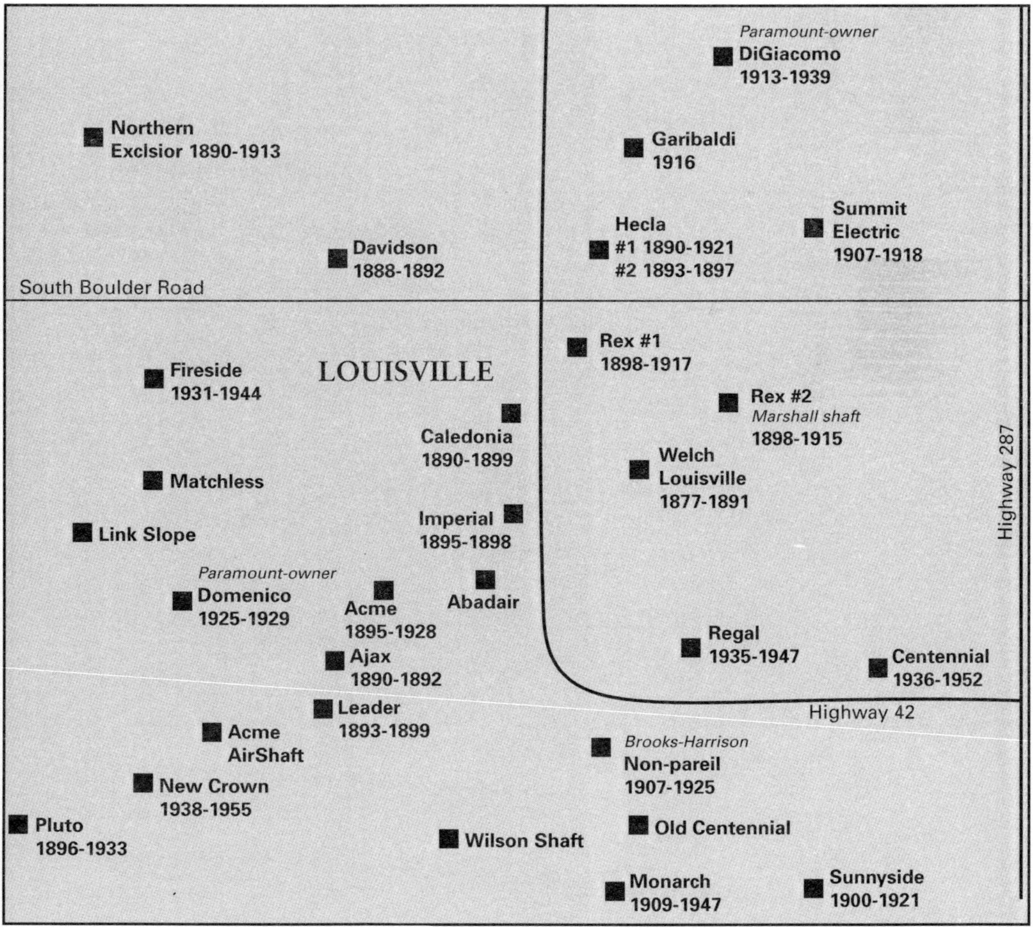

Lafayette

Baker	1887 - 1896	Lincoln (Old)	1896 - 1902
Baker New Shaft	1905 - 1909	Mile High	1913 - 1918
Baseline	1941 - 1942	Mitchell	1893 - 1897
Black Diamond #2	1931 - 1942	Monroe	1917 - 1925
Blue Ribbon	1905	Munroe (Baseline)	1933 - 1940
Blue Goose	1921 - 1924	Parkdale	1916
Cambro	1917 - 1928	Senator	1890 - 1913
Canon	1888 - 1898	Simpson/Spencer	1888 - 1926
Cannon Shaft (Otis)	No date	Standard	1887 - 1926
Capitol	1908 - 1926	Storrs	1901 - 1904
Caryl Pallott	1898 - 1900	Strathmore	1901 - 1919
Gladstone	1890 - 1915	Vaughn	1897 - 1906
Haywood	1899 - 1908	Willoughby	1908 - 1913
Irvington	1907 - 1908		

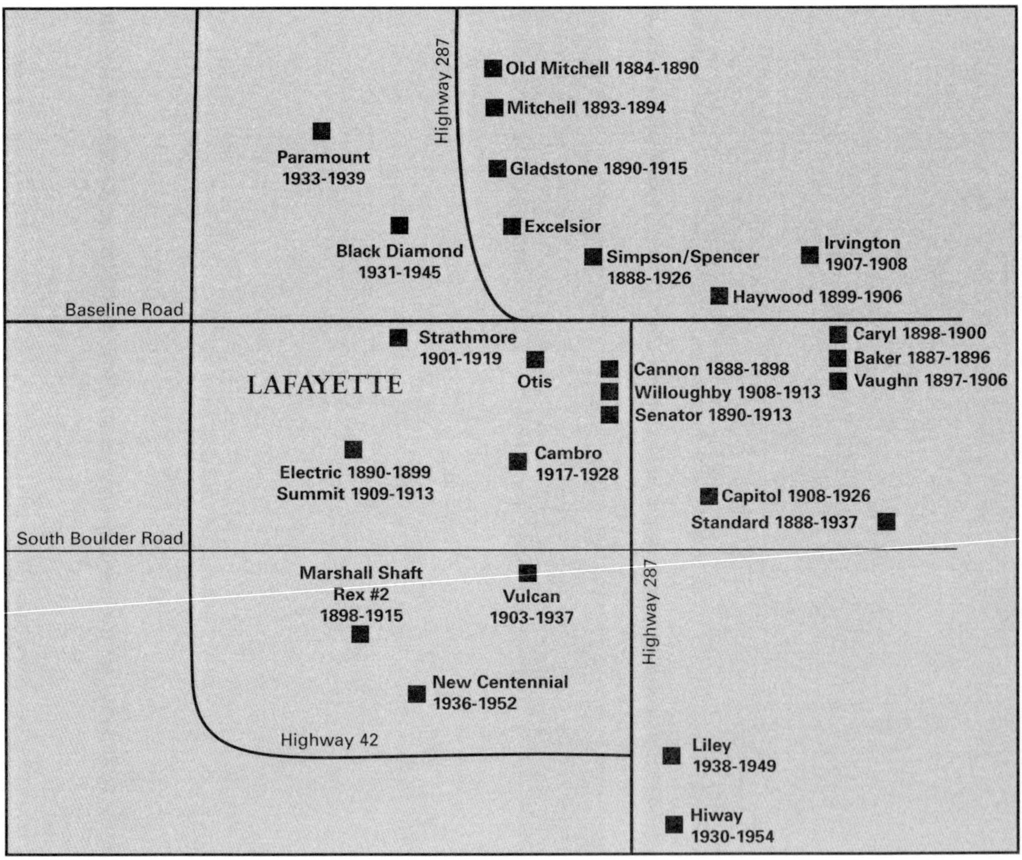

Erie and Candfield

Mine	Years
Arrow	1933 - 1935
Banner Shaft	1932 - 1934
Boulder Valley #1	1917 - 1947
Boulder Valley (Old)	1883 - 1885
Briggs Shaft	No date
Canfield Shaft	1927
Chase	1892 - 1895
Clayton	1920 - 1942
Cleveland	1885 - 1895
Columbine	1920 - 1946
Cresent	1905
Eagle	1933 -1975
Erie Strip	1948 - 1953
Eureka #1	1900 - 1926
Eureka #2	1900 - 1926
Eureka #3	1900 - 1926
Garfield #1	1884 - 1897
Garfield #2	1886 - 1905
Ideal	1909 - 1916
Imperial	1927 - 1972
Jackson	1885 - 1890
Lehigh	1902 - 1910
Lloyd Shaft	No date
Wooley	1897 - 1898
Longs Peak	1892 - 1901
Marfel Shaft	1897 - 1904
McGregor	1884 - 1895
Mitchell (New)	1893 - 1897
Morrison	1930 - 1953
Northrup	1883 - 1884
Northwestern	1883 - 1907
Puritan	1908 - 1939
Reliance	1903 - 1907
Shamrock	1905 - 1956
Standard	1888 - 1937
Star (Old)	1883 - 1895
Star (New)	1918 - 1921
State	1913 - 1916
Stewart	1883 - 1892
Tynon	1900 - 1904
Washington (Old)	1903 - 1911
Washington (New)	1916 - 1918
Washington Shaft	No date
Washington	1940 - 1967
Whitehouse #1	1896 - 1907
Whitehouse #2	1896 - 1907

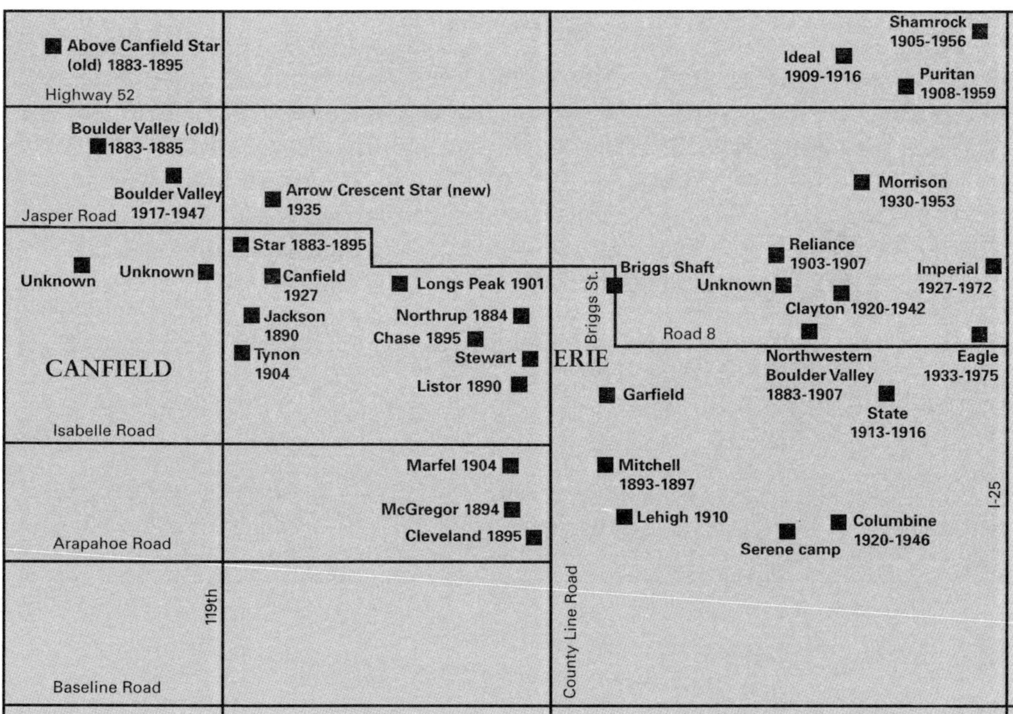

Firestone

Davis	1900 - 1906
Emmerson	1897 - 1904
Evan Jones Shaft	No date
Grant	1914 - 1939
McKissic	1887 - 1907
Russell	1914 - 1947
St. Vrain	1890 - 1892

Frederick

Baum	1914 - 1953
Evans	1907 - 1942
Firestone	1908 - 1920
Frederick Slope	1907 - 1929
Golden Ash	1906 - 1913
Warwick	1908 - 1912

Dacono

Boulder Valley #3	1944 - 1969
Graden	1935 - 1955
Hingley Morgan Shaft	No date
Lincoln	1948 - 1974
Lincoln Shaft	No date
Peerless	1920 - 1965
Witherbee	1920 - 1937

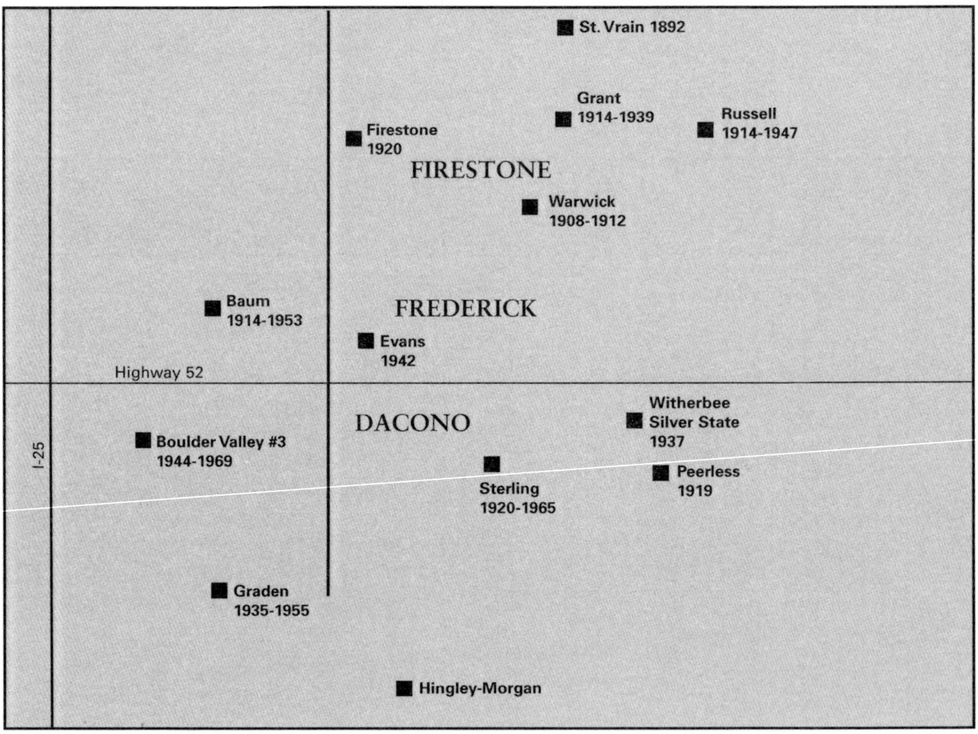

Mining Methods

The Northern Field had slope mines, usually cut into sides of hills where a coal vein rose to the surface. The majority of mines were deep mines, however, opened when a shaft was sunk.

Digging the shaft was the only way to open the mine. The first and early mines were opened years before machines were available to do the work in the Northern Field. Therefore, the digging was manual labor performed by teams of two men working with only pick and shovel. As the dirt was removed the sides were stablized with cribbing that became the wall of the shaft. The dirt from the shaft was taken to the surface by the bucket load. One hundred, two hundred, three hundred feet later, when the team had reached the layer of coal that was to be mined, the shaft was opened, lined and ready for the mechanized lift that brought the miners and their car loads of coal to the surface.

Knowing where to dig some of the early and smaller mines was as unscientific as the work was unmechanized. If the coal wasn't visible on the surface or in a hillside, the digging to open a mine might be only speculation. By the time large coal companies were organized to open the mines, core samples were drilled to figure location and depth of the vein so opening of the mine could be more orderly.

Once the mines were opened, work to remove the coal began. The mining method used in the deep underground mines was the room and pillar system. The main route through the mine was the haulageway with tracks laid for small coal cars so the coal could be removed from the area. Leading from the haulageway was the room from which the coal could be cut. The room neck was the entry with a pillar of coal at least ten feet wide left in place to support the ceiling. Layers of coal were left at the ceiling and sides of the room to support the earth surrounding the room.

An air passageway was alongside the haulageway to provide air circulation and ventilation. Brattice cloths were hung in passageways to control the air flow. In the case of a fire in the mine the cloths served as an emergency wall to contain the fire. The wall could then be plastered over to seal off the room and the fire.

Mine tunnels were needed only when cuts were made through a rock fault to develop another section of the mine.

Miners worked in pairs in the rooms with the miner's very life depending on the partner's skill. The first coal in the room was cut out by hand. In the widest veins the miners could stand for this work. Miners had to work in a prone position in the narrowest veins.

To make the coal available for removal an explosive charge was used to loosen the supply and to enlarge the room. Miners prepared a charge of black powder which was pushed by a tampen to the back of the hole that had been drilled into the coal. The charge was plugged so it would fire into the coal and not back out. The shot was the fuse to ignite the charge to produce the small explosion to loosen the coal.

Punching machine on exhibit in downtown Louisville was rescued from a closed mine by former State Mine Inspector Andy Deborski. Punching machines dated from 1880's.

Air-powered punching machines were developed to do the initial cutting of the coal and were later replaced by cutting machines.

A cutting machine developed in the 1940's by a miner at the Baum Mine at Frederick was the Continuous Miner, a machine that could cut away the coal from the room, load the coal onto a conveyor belt and then into the coal cars to be transported to the surface. The idea for the machine was given to a manufacturing firm in Denver to produce. The electrically powered machine was so ingenious that representatives from mines around the world came to the Baum Mine to investigate the new device. The Continuous Miner not only reduced production costs but the mining process was safer because explosives were no longer needed to loosen coal for easier access.

Other skills were needed to liberate the coal:

Timbermen sawed logs which were used as props to brace ceilings and augment the coal pillars.

Diggers were needed to drill holes for the shot which detonated to release the coal so it could be picked away and loaded onto the car.

The shot firers were a specialized group and their skills were critical to the process in the mines which often contained dangerous gases.

The fire boss prepared for his work through study and testing. He entered the mine before each shift to examine for gas. His equipment was a safety lamp. The lamp's enclosed flame would be extinguished if black damp, carbon dioxide gas, was present. The flame would flare if an explosive gas was in the mine. In both cases the miners needed to leave the mine immediately. [9]

Miners, or diggers, were paid by the amount of coal they could shovel into a car. The cars were then moved to the lift and from there raised to ground level and sent to the mine's weigh station.

Day workers, or those paid by the day, were the track layers, timbermen, pumpmen, electricians and blacksmiths.Work that had to be done with no pay scale included debris clean up, loading of the coal, and removal of unwanted debris from the coal.

Working with the crews were the mules which were used to move the coal cars from rooms to the lift. The mules were housed inside the mines and most never left. They were named, were well treated and were considered to be intelligent animals. The mules were used beyond a time of mechanization in those mines which had gas, since battery powered motor cars could create dangerous sparks.

The heavy cars were dangerous and with no hand brakes on the cars, they were difficult to stop once in motion. The mules could be trained to step off the tracks to stop the load but most often the men depended on throwing a sprag, made of metal, into the spoke of the wheel to slow down the "trip" of cars. Lack of space in the haulageway made the work dangerous.

When a mine was worked out and closed, the pillars were pulled and the room collapsed. If the room was far enough underground the collapsing earth would fill the void and leave the surface undisturbed. In some cases the pillars were not collapsed and years later the timbers would deteriorate and give way. If the collapse did reach the surface there would be a sudden drop in the ground level, or subsidence.

Longwall mining was a method of mining used in Great Britain and Europe but not used in Northern Field. The long wall of coal was removed as the mining progressed allowing the roof to sag into the void with surface subsidence more likely to occur at that time.

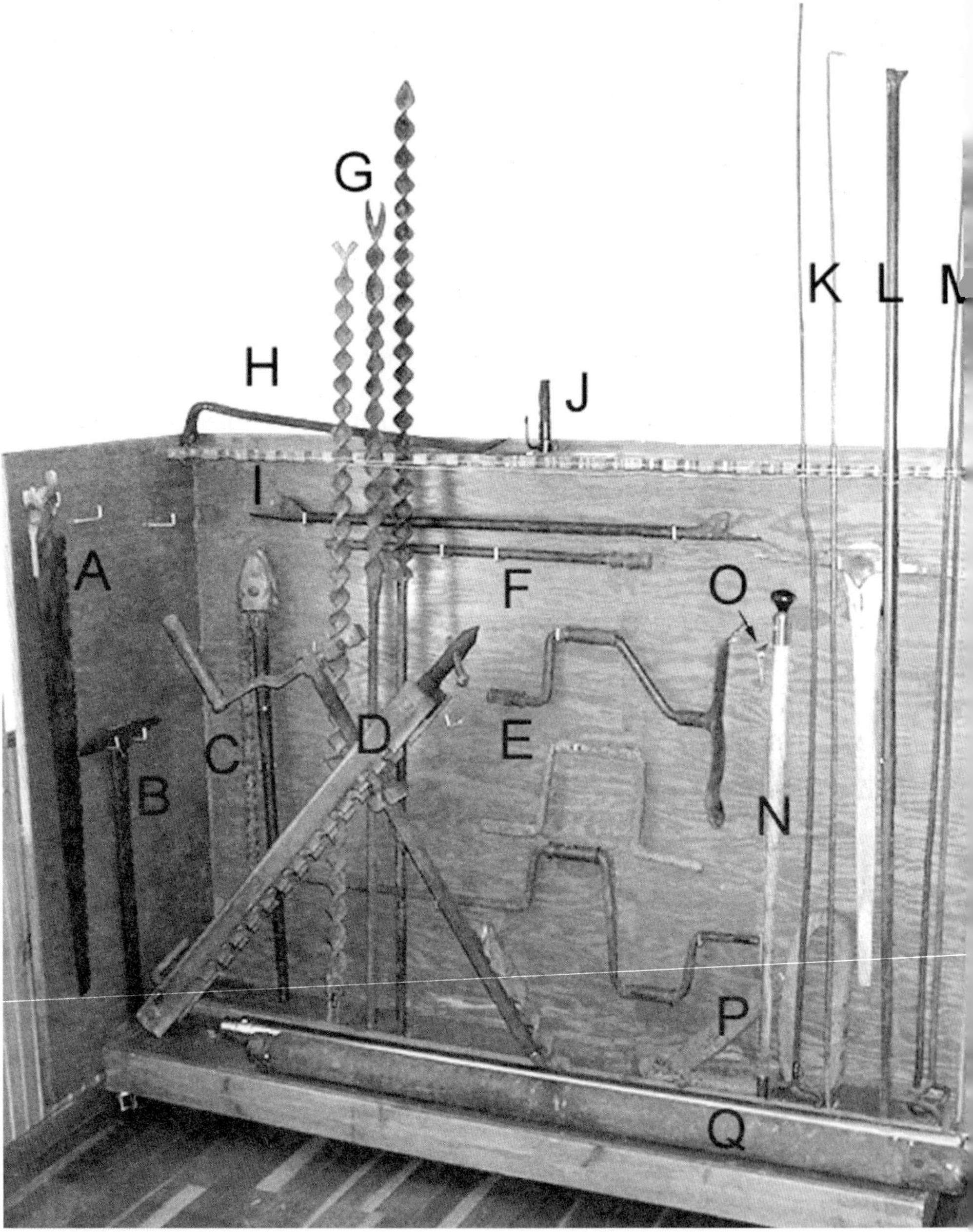

Exhibit of tools for mining of coal and explanations of uses of the tools on display at Louisville Historical Museum. Research and display prepared by Carol Gleeson, Museum Director, and Todd Gleeson. Original organization of work tools by Betty Buffo with help from Don ross and andy Deborski.

Equipment

A. TIMBER SAW. An example of a metal handled saw used to saw timbers in the mine. Early miners were not paid for their time spent timbering.

B. PICKS. Picks of various sizes and shapes were used depending on the hardness of the coal and the location of the seam. This display has two examples.

C. SPANNER WRENCH. This wrench is an example of tools used to assemble and disassemble plumbing within the mine.

D. JACK STAND & AUGER. The jack stand (a.k.a. thread post) could be wedged tight between the floor and roof of a coal vein. A thread bar was then mounted into the jack stand and augers of various lengths attached at the end of the thread bar and used to drill a bore hole for blasting. Note that the jack stand is relatively short. Northern coal field veins were narrow, often about 3 feet thick, which forced miners to work on their knees or side.

E. CRANK AUGERS. These drilling tools were used for boring shot holes in coal in preparation for blasting. One of the three examples has a BREAST PLATE attached, which allowed a miner to press the auger forward with his body while drilling.

F. AUGER EXTENSION. Extensions were used with hand powered augers to drill extra deep holes.

G. WORM AUGER BITS. Worm refers to the spiral shape of the auger tip. "Worms" were used to loosen and drill through the tough clay to create a bore hole for blasting. CASINGS were sometimes used to line the inside of a bore hole as it was drilled, keeping out water or preventing collapse.

H. PRY BAR. Pry bars were used within the mine to dislodge rock, set or remove timbers, and perform general functions.

I. SPIKE BAR. Spike Bars were heavy steel rods with pointed or flattened ends, and were used to pry or dislodge rock.

J. MINER'S CANDLE STICK. Candles were mounted in or hung from a timber or rock crevice and used to light a tunnel. The open flames of candles and lard oil lamps were largely replaced by carbide lamps around 1915.

K. SPOONS. Spoons were used to clean out a shot hole prior to charging the hole with black powder (a mixture of potassium nitrate, powdered charcoal, and sulphur) for blasting.

L. TAMPING BAR. A tamping bar was used to pack the shot hole with clay or rolled newspaper after charging the hole with powder. The groove in the tamping end was used to follow the COPPER NEEDLE (M) to the powder charge while keeping the needle in place.

M. COPPER NEEDLES. Needles were long, narrow rods of copper (or sometimes brass) tapered to a point at one end and turned into a handle at the other. They were used to position a rolled paper cartridge of black powder into the shot hole. The hole was then tamped and plugged with clay or newspaper around the needle. The needle was then withdrawn carefully, leaving a straight passageway through the tamping for the miner's SQUIB. A SQUIB was a thin tube filled with black powder, forming a slow (~30 sec) burning fuse to "shoot" or fire the charge. Copper or brass needles were used to prevent sparking.

N. COAL ROOF TESTER (A.K.A. SOUNDING BAR). A Fire Boss or Foreman tapped the roof of the mine with a roof tester to determine if it was solid and safe. It has a copper tip to prevent sparks.

O. SPADS These small spikes were used to mark coal veins or to guide a surveyor's line within the mine. Some spads were made of horseshoe nails with a hole drilled through its head.

P. COAL CAR LINKS AND HITCH PIN. These were part of the linkage system used to fasten multiple coal cars together.

Q. CARDOX SHELLS. Steel shells charged with compressed carbon dioxide and used to break up the coal seams. They had an electric heater in one end. When electric power was applied to the heater, the gas expanded and increased pressure which ruptured a disc in the shell and released the high pressure gas. The Cardox method of blasting was seventy-five times slower than black powder which resulted in larger pieces of coal and less slack (very small pieces of coal). Cardox blasting techniques came into use in the 1930's.

Coal cars in a trip moved coal from mine room to lift and then to surface. Mules pulled the cars through the mine on tracks.

Hecla Mine room with timber pillars. The mine opened at 180 feet in 1890. Mined on two seams, production was at 1,427,000 tons until 1920. Rocky Mountain Fuel Co. was the last operator.
Photo from collection of Louisville Public Library and Louisville Historical Museum.

Mine Lighting

The lighting system for the miners began in a way that was as primitive as the mining methods had been. Necessity prompts invention and improvements were made to systems throughout the years of mining. In each case, the newer systems were somewhat safer than the old way.

Wrought iron candle holder

First illumination in mines was a candle pounded into a support timer. The open flame was a danger in mines where coal dust and natural gas were problems.

Oil or pot lamp
The oil lamp used solid parafin, melted from the heat of the flame, with an open flame for the light. It was considered an improvement as the lamp was worn on a cap by the miner and moved with him where light was needed. As it was also unsafe, the lamp was replaced by the carbide lamp and finally by the battery-powered lamp.

A safety lamp was carried into the mine by the fire boss to detect presence of gasses or absence of oxygen.

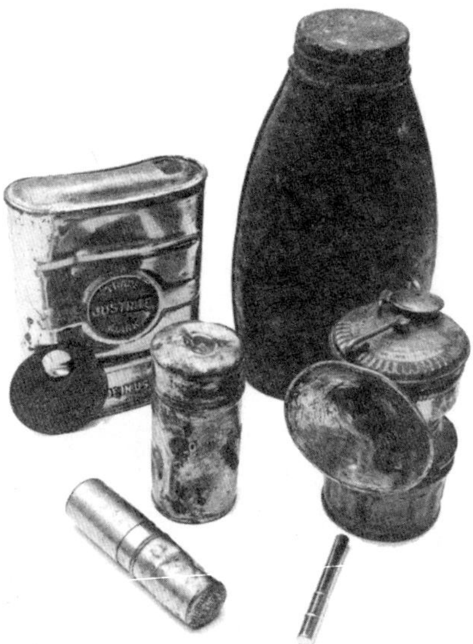

Carbide lamp and equipment
The carbide lamp became standard equipment after 1914 and was considered safer although there was still an open flame. Water from the upper chamber dripped onto carbide forming a gas which could be ignited. Miners carried extra carbide into the mine for a continous light during the work day.

Mining Caps

Canvas miner's cap, on left, was the first type of head gear worn in mines. Hard hats became required equipment in 1920's. The oil lamp was declared unsafe soon after that.

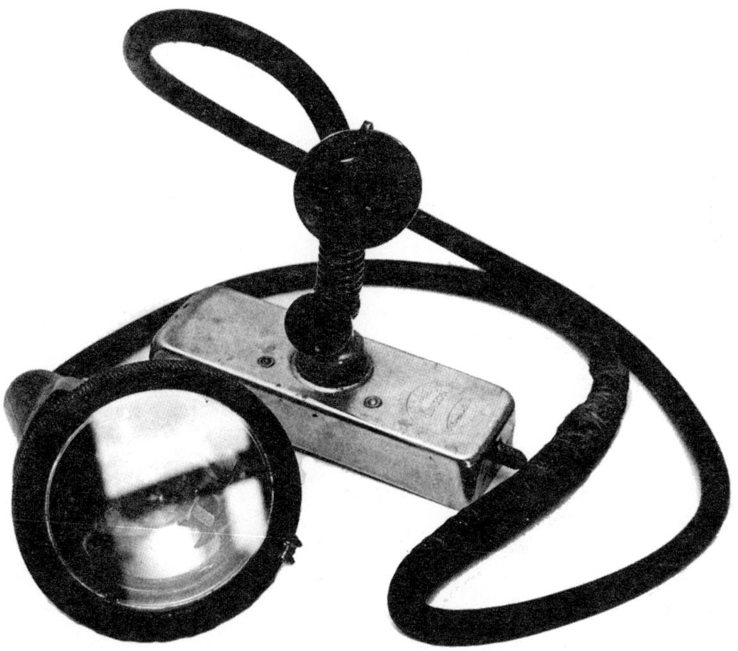

Battery powered lamp

Battery-powered lamps for miners' safety caps brought lighting efficiency and safer conditions for their work environment.

Accidents

Mining was hard work for long hours. The early mining was done without the benefit of mechanized, labor-saving equipment. Men worked with skill, not fear, although there were dangers.

Mine accidents were recorded in ledgers in offices at the mines. After 1909 when Colorado Governor Shaffroth ordered inspections in the coal mines, the accidents became a matter of state record. Injuries and deaths were statistics, a consequence of working with the danger. A fall of coal or rock from the ceiling of a room caused many deaths and injuries. Coal cars in motion had to be avoided. Coal dust and gas caused explosions.

Accidents reported from 1900 to 1945 in Erie and Frederick districts: falling coal and rock, 9; coal car collisions, 6; falling into shaft, 4; miscellaneous, 3. [(10)]

For decades operators held miners responsible for accidents: they had been careless. An inquest into a fatal accident would rule the cause as unavoidable and the operator had no liability.

Explosions were a danger to which miners were always alert. Newspaper reports in 1902 told of two explosions. Sunnyside Mine at Louisville had four gas explosions in one month. The last one, April 24, killed one miner and another miner had been badly burned.

The Simpson Mine at Lafayette also experienced explosion and fire in 1902. The Denver Times reported that gas exploded November 20 at the Simpson and fire followed. Men and mules were evacuated to the surface. Air was turned to force the fire toward the shaft and water was then poured on to the fire. Men and mules were back at work by November 24.

The worst mine explosion in the Northern Field was at Monarch Mine, south of Louisville, in January 1936. Eight men entered the mine to do the daily work preparations before the day shift workers entered. The fire boss was inspecting because coal dust had been accumulating on the mine floor. The force of the explosion shattered mine timbers and rock falls filled the mine in an area where two passageways came together. It was speculated that two coal cars had collided at the intersection during the pre-shift work and sparks from the collision ignited the coal dust.

Two men in the mine at the time of the explosion escaped, Nick DelPizzo and William Jenkins Jr. They were able to reach the air shaft and climbed 300 feet to safety. Steve Davis, the fire boss, died from the after damp gas. [(11)] Seven men, Ray Bailey, Oscar Baird, Tom Stevens, Tony DeSantis, Kester Novinger and Leland Ward, died in the explosion. Joe Jaramillo, who was the mine's mule driver, died at the center of the rock fall and his body was never recovered. Jaramillo's monument, which had been placed over the approximate site of the explosion, was moved when Flatirons Crossing Shopping Mall was constructed. The marker is now in a park area on the north side of the mall.

Monarch Mine was reopened for mining as soon as the debris could be cleared away. As in every case of explosion, fire or mine accident, whether fatal or not, the miners returned to their work at the first signal of their mine's whistle.

Chapter 4
Miners Begin to Organize

Wages

Considering the difficulty of the work and the skill required, the issues of safety in the mines and seasonal nature of the work, wages for the coal miners in the Northern Field were always under negotiation.

Northern Field coal would dry out when brought to the surface and then would crumble if held in storage. This quality dictated the operating schedule for the mines. The Denver market for coal was predominately for heating. Winter was good for the coal field. Summer work was minimal.

Lafayette Leader, newspaper of record for Lafayette until 1970, was considered to be friendly to the union cause and often ran reports of wages earned. The articles coincided with times of union strike activities and served to inform readers of the need for the requested wage increases.

1878 — Pay at the Welch Mine in Louisville was $2.50 a day because it was a new mine and considered safer. Pay at the Northern Field's older mines was $3.00 to $3.50.

1881 — $3.50 to $4.50 a day.

1901 — Mitchell Mine reported for the work of 20 men for the month's 21 days of work:

Coal mined - 3,047 tons - 152 tons per man
Average wage - $38 at 25c a ton
Expenses deducted: doctor, $1; squib, 25c; oil, $1.20; powder, $6.25; blacksmith, 50c; check weighman, $1.53.

1913 — $3.50 a day; 174 days of work - annual income, $696.
Expenses deducted: blacksmith, explosives, medical expense

1930's — $4.25 a day, 32c a ton.

Lafayette Leader reports of salaries in the community in 1914:
School District report for July 1913 to June 1914 — Total salaries paid to teachers (number not reported) — $9,977.
Town of Lafayette, bills paid for April 1914 include:
S. R. Wood, salary $8.00; C. A. Magnuson, clerk, salary, $60.00; Total April bills, $2,033; Newspaper subscription for year, $1.50.

In days before union acceptance, miners paid for the supplies that were used in the work of mining. Charges were figured each month for the squibs and fuses, as well as the powder, used in dynamiting the coal from the walls of the room. A blacksmith was on hand in the yard to repair equipment, but a monthly fee was charged for the service. The check weighman was in charge of the scale to weigh coal production. He was not popular with the miners and his decisions about the weight of the production were seldom satisfactory to the miners. Nevertheless, his decisions were final and a deduction was taken from the miners' pay envelopes for the weighing service. Another fee paid by the miners was the medical insurance payment although it was not called insurance. The company's designated doctor received $1 a month, deducted from each pay envelope, for all medical services for the miners and their families.

A young boy could begin to work in the mine if he could work alongside his father or other relative. By assisting and watching he would learn the skills his father could teach him and children were known to enter the world of the mines by the age of eight. Even in the later years of the 1930's, boys often completed their formal education by the end of the 8th grade.

Towns in the Northern Field were not company towns, and the miners did not shop in company stores. To survive the summer without work the miners had big gardens, worked at any job available and depended on a good credit record at their local grocery store. Purchases on credit were always paid in full when mines operated again, even if a relative had to help out in order to save the family name. Customers depended on the grocer's good will during the winter-summer work cycle, and the businessmen were essential to the town's economy.

Unions

The miners did demand a working environment that was as safe as could be made. They did ask for a wage that would pay for a modest, but adequate, living for their families. Their struggle for these goals covered the years of the industry in the Northern Field. They realized early on that by working together in asking for their safe environment and fair wage they might be able to have more success. However, it was many years before the goal of union representation and collective bargaining could be achieved. There were many strikes, work stoppages, and hard times along the way to achieving this goal.

Knights of Labor was the first national labor organization in the United States, and the first meeting was held in 1869 in Philadelphia.

It was only five years later, in 1874, that the first Knights of Labor union in Colorado was organized by local miners in Erie.

With the coal mines opening in the Northern Field, the miners arrived. They were eager to find jobs in the mines - happy for the pay, happy for the opportunity. They were skilled and hard working, but pay was minimal for the long working day and dangerous conditions. When Eastern miners began to move to the West bringing the union message with them, the Erie miners were ready to listen.

The long 10 to 12-hour work days, the seasonal unemployment and the hefty deductions taken from pay envelopes began to make the idea of the larger group asking for some relief seem like a very good idea to some.

The first strike was in 1871 and working life in the Northern Field was disrupted by a strike every few years for the next 70 years.

Higher pay and an eight-hour day were the primary issues year after year. Eventually the safety of the mine became a government concern and laws were passed to bring standards of safety to the mines. But always the miners wanted the union recognition.

Two union organizers for Knights of Labor arrived in Erie to foment the 1871 strike. A riot between strikers and non-strikers resulted in arrest and fines for the leaders.

By 1878 Knights of Labor union was strong enough to register Local #771, and the Erie union had 60 members. All mines in the Northern Field had a union-type group by 1880, but there was no organization between the groups. Louisville's union was registered as Coal Miners Federation.

Coal miners gained a powerful voice in 1890 when United Mine Workers of America was founded in Columbus, Ohio. Knights of Labor merged with the National Progressive Union of Coal Miners to form UMWA.[12] Erie's Knights of Labor local became the first UMWA local in Colorado.

The first activity for UMWA in Colorado came in 1903 when Mother Jones was sent to organize a strike to improve conditions in the Southern Coal Field. Only a small percentage of miners in the Northern Field were members of UMWA locals, but the union called the strike which closed mines and lasted three weeks.

From the 1903 strike until the miners' work slowed down and mines began to close in 1950's, there was unrest in the Northern Field. The Wobblies, International Workers of the World, is the union given credit for organizing the 1927 strike that led to the conflict at the Columbine Mine in Erie. But it was John L. Lewis, from Eastern headquarters of UMWA, who determined the miners' strikes and work schedules through the years of mining.

Salary increases, mine safety, and shorter work days were gradually achieved through the unrest and strikes, but recognition of the unions was refused by owners. The federal government intervened in the form of the National Labor Relations Act of 1933 to give the miners their right to collective bargaining. But union activity and the strikes were part of life in the Northern Field.

Underground at Eagle Mine, Erie
EAGLE MINE EAST OF ERIE CLOSED IN 1975 AND WORKED FOR 38 YEARS. PRODUCTION WAS 7,758,500 TONS OF COAL AND SHAFT DEPTH WAS 370 FEET. EAGLE MINE WAS OPERATED BY IMPERIAL COAL CO. AND WAS THE LAST MINE TO WORK IN THE NORTHERN FIELD. EVERETT WEST IS THE MINER IN THE PHOTOGRAPH.
Photo from collection of Lafayette Miners Museum

Chapter 5

Conflicts in Northern Field

The mines had opened and the miners came to work. The union organizers and their calls to strike were soon to follow, but miners didn't need union organizers to tell them of the hardships and dangers of their work. They didn't need to be told that a refusal to work, a strike, might bring improvements in mining conditions.

The first strike at one of the first mines in Northern Field was in February 1871 at the Boulder Valley Mine in Erie. A strike was called again in 1875, again at the Boulder Valley Mine, when the operator tried to reduce wages. An attempt was made to hire a new group of workers to replace the striking workers but the union cause prevailed and the company was forced to back down.

Marshall miners went out on strike in December 1879 demanding higher wages, and the miners won their increase. Erie and Canfield miners followed with their own strike but this time strike breakers were hired and the Erie/Canfield miners lost ground. (13)

Between 1880 and 1904 there were 13 strikes in area mines. (14) Knights of Labor union organizers were successful in bringing cooperation between miners and companies during the period between 1886 and 1890 when no strikes were called. With the organization of United Mine Workers of America in 1890, the separate unions in the communities of Northern Field began coordinating their efforts - and their strikes.

One miner from the Mitchell Mine in Lafayette wrote an account of grievances for the Rocky Mountain News. The miner explained for Denver readers why there might be a coal shortage in the winter of 1901. The article was reprinted in Lafayette Leader on January 14, 1901.

He said a shortage of workers was not the problem as men were shipped in every day to work in mines. Untrained workers often took away jobs from the miners.

Another problem was the weighing process. Two miners would "blast coal down" (at their own expense for the supplies) and load up 20 tons of coal. The cars would be run over "a small platform scale to be weighed on the fly without stopping and the 20 tons of coal the two men had sent up had shrunk to 15 tons in the company's hands."

At that rate the two men made $3.75 for the day's labor.

The miner figured that 12 years ago miners earned $1.25 per ton and the mine operator, this case the Northern Coal Co., charged $4 a ton in Denver. In 1901 miners were paid 25 cents a ton and owners charged $4 a ton in Denver.

He said "The miners are not content the way they are dealt with."

The law of supply and demand in the coal mining industry affected the mines and towns in the Northern Field. Government regulations did not control mining operations or the opening of new mines.

Big companies with absentee ownership, small local companies and independent miners opened mines. A group of 18 miners who were out on strike in 1893 formed the Cooperative Mining Co. They were able to open their own mine, without capital, and provide the labor in their own mine for several years.

The seasonal demand for coal and the competition from the many mines created the problem of excess supply. Soon there were too many miners as the mine owners would bring in new miners to keep the mines in operation during times of strikes. Some were brought in from Europe and some, as the Mitchell miners said, were untrained workers.

Mine owners reduced wages to be able to meet expenses of operation and make a profit. And they continued to deny union recognition and collective bargaining.

Northern Field miners were organized with UMWA locals in all communities. The union reported a membership of 11,000 miners in Colorado, about 15% of the state's miners, many of them in Boulder and Weld counties.

As mines began to age they became unsafe. With the oversupply of workers, the work was sometimes scarce. Times were difficult, but the Northern Field miners lived in independent communities and owned their own homes. The mine owners lived in the area, at least no farther than Denver. Owners were not from the East coast.

Life in the Southern Coal Field of Colorado was not so democratic. The Southern Field centered around Walsenburg, Trinidad, and at mines located into the foothills of the Rocky Mountains. The mines in Las Animas and Huerfano counties produced 60% of the coal in Colorado at the time of the first union attempt to organize. [15]

Communities at the mines were company camps with primitive living conditions. Rigid control was maintained for the absentee owners by guards and foremen hired to oversee the camps. Miners were trapped. Union activity was not allowed and miners were without recourse, until Mother Jones came to Colorado to help the miners' cause.

Mother Jones

Mary Harris Jones lived a quiet, normal life until her husband and four children died in an epidemic (yellow fever has been reported) in Memphis, Tenn. She was 60 years old.

The labor unions were just beginning to organize. Mrs. Jones was so disturbed by the working conditions in the southern cotton mills where many children were working that she joined the fight for labor's rights and benefits. She was white haired, well rounded and with her pleasant expression she became known as Mother Jones. She was mild looking but she had the determination of a tiger. Her labor concerns took her to cotton mills, steel mills, the pullman strike and the coal mines of Pennsylvania.

The UMWA asked her to go to Colorado in 1903 to organize strikes for the coal miners, and Mother Jones was ready for the challenge.

Mother Jones is reported to have attended the District 15 meeting in Louisville on Nov. 21, 1903. She had gone first to the Southern Field and was horrified at working and living conditions there. She wanted all the mines in the state to be on strike to enable the Southern Field miners to overcome the power of the mine owners.

The mission at the Louisville meeting was to convince the union miners to turn down the owners' offer to end the strike and in so doing to support the Southern Field strike. Louisville union miners voted to support Mother Jones' request, but the non-union miners were outraged that the strike would continue. They talked the union miners into taking a second vote and the Northern Field strike came to an end.

Mother Jones was angered and returned to the Southern Field to encourage the miners to continue their fight. Her actions led to a jail term in the Trinidad jail. The UMWA had sent nearly a half million dollars to Colorado to support the union strikers and headquarters was not pleased with Mother Jones' Southern Field activities or her failure to hold the union together. The UMWA ordered her out of Colorado, and the union and Mother Jones began to part ways.

She returned to Colorado in December 1913, during the height of the Southern Field strike. By this time she was 82 years of age and her very presence gave hope to the union members and their cause. At an appearance in Denver umbrellas were lowered to give her honor, according to a newspaper report. Southern Field operators were not so pleased to have her talking to the strikers. She was held in jail for nine weeks by the militia, and when she was released she was ordered deported from Trinidad and Colorado in January 1914.

Mother Jones was no longer part of the United Mine Workers of America and had begun to move toward the ideals of the Socialist party and the International Workers of the World. [16]

The 1903 strike had been a mistake. The union was not well enough organized to work together and Mother Jones could not promote the cause or control the members.

The Northern Field miners achieved some success, a 15% wage increase and the eight-hour day. Mines in the Southern Field had continued to produce as usual. Owners refused to even recognize the strike, let alone a union, and the miners continued to work in the terrible conditions.

Northern Field mines continued operation but with less production and union activity declined for a few years after 1903 although UMWA organizers did not go away. Mine owners continued to be in control of the industry while miners worked with growing frustration at conditions and wages.

A Northern Field strike was averted in July 1908 when mine operators signed a contract with miners, the first major contract for Colorado miners. UMWA representative John Lawson worked with Lafayette miners to negotiate the contract.

Miners gained an eight-hour day and improvements in mine safety, but the most significant achievement was formation of a grievance committee to help settle disputes. The union had finally achieved recognition by at least one Colorado mine operator. (17)

John Shaffroth, Democratic governor of Colorado from 1909 to 1913, promoted legislation which gave support to the union. Shaffroth ordered the eight-hour day for all workers and called for coal mine inspections for mine safety.

Northern Field miners had to be satisfied with the improved working conditions won by the union in the contract negotiations of 1908. However, their contract for their work expired on April 1, 1910.

Strike of 1910 - 1914

The miners asked for a 15 percent wage increase and owners refused saying Northern Field miners already earned 20 percent more than Southern Field miners. Operators said they could not compete with Southern Field coal coming to the Denver market with the rising costs for coal production.

By April 3, 1910, all mines in the Northern Field were closed. On April 4 operators met and signed an agreement saying the mines would now become "open mines" and would be opening soon. In order to reopen miners would be brought in who would go to work. The imported strikebreakers, called scabs, had the mines operating by April 19.

The operators had to build company housing for the non-union miners and familes and the houses were enclosed in a fenced compound. The Hecla Mine operator, the Citizen Coal Co., hired guards to protect the workers.

American Fuel Co. operated eight mines in the Northern Field: Matchless, Centennial and Electric Mines at Louisville; Capitol and Senator mines at Lafayette; State Mine at Erie; Fox Mine at Marshall; Evans Mine at Frederick.

Northern Coal and Coke Co. operated the Acme at Louisville; Industrial at Superior; Simpson and Vulcan mines at Lafayette.

In October 1911 Rocky Mountain Fuel Co. inherited the angry strikers and the strike breakers when it acquired the properties of the Northern Coal and Coke Co.

and the Citizen Coal Co. mine, the Hecla.

Over 2,700 miners in the Northern Field settled into the longest strike in coal mining history. Confrontations began between strikers, strike breakers and hired guards. The governor responded to requests by the mine operators in the Northern Field, and later in the Southern Field, and sent in state militia. The plan was to restore order and peace to the coal fields, but the militia's presence only increased the confrontations. There were four long years of anger and conflict before the federal troops were ordered to Louisville and other towns in Northern Field. The troops arrived in April 1914 to bring peace to the mines and their towns.

An allotment came to union members from the UMWA of $1.50 a week for adults and 75 cents for a child during the strike years to give some relief. However, the strike was a bitter experience and never forgotten by those who lived through it.

American Fuel Co. mines, from Marshall to Frederick, were willing to negotiate and settled strike grievances with the miners in March 1912.

District 15 UMWA officers and John Lawson, who was now a member of the Board of Directors of the UMWA, negotiated the agreement and 1000 men returned to the mines. They had earned a 3 cents a ton increase and 5 cents more for the day work.

The settlement ratified an agreement signed in 1908 for a uniform pay scale. Known as the Louisville Scale, it was a calculation for pay for loading coal "after machines" or "pick work." There also were calculations for working in coal veins over six feet (which meant that a miner could stand upright while he worked) or under six feet.

The UMWA Board member and local organizer, John Lawson, worked in the coal mines while he talked to the miners about union membership. He helped to organize the strike but before the strike was settled he was working to prevent violence.

John Lawson

Lawson was born in Scotland and began to work in coal mines in Pennsylvania as a chiild. His family moved to Colorado and he continued mining as an assistant to his father. He joined the Knights of Labor and was soon talking with miners about the importance of unions.

Lawson became an organizer for United Mine Workers of America after the group was formed in 1890 and in 1907 he was appointed to the national UMWA Board of Directors. He was sent to Trinidad, Colorado, to begin again the union organizing effort in the Southern Field where operators had total control of mines, housing and lives of the miners. Lawson's union activities and organizing efforts were not welcomed by the operators.

Lawson traveled back and forth between Northern and Southern Fields to be where he was most needed as events unfolded. He was in the north to settle the strike in 1908 and he returned to negotiate with American Fuel Co. operators in the March 1912 agreement.

By September 24, 1913, Southern Field miners had responded to Lawson's organizing efforts and gone out on strike. Southern Field mines were closed down.

The two years of strike activity in the Southern Field were violent as miners were enraged over the deaths that were occurring on the streets of Trinidad and around the mines and the mine camps. The miners, many who were foreigners and not English speaking, wanted revenge. Lawson worked tirelessly to calm the miners and to prevent more violence but both the Northern and Southern fields erupted into open warfare as mine guards fired on striking miners. The tragedy of the strike unfolded as Lawson did what little he could to bring matters to a close.

During the turmoil of the shootings and fires in the Southern Field he evacuated women and children from the Ludlow tent colony, he moved among the miners who were firing at the guards, and he searched through the tents trying to rescue any who had been left behind after fires were started.

The strike ended in December 1914 and in May 1915 Lawson was on trial for his life. He was falsely charged with murder plus bringing in weapons and directing the miners' fight in the Southern Field. He was first found guilty but eventually his conviction was overturned.

The UMWA abandoned Lawson and in February 1915 he was fired from his job with the union. He returned to mining and worked at Lafayette's Ideal Mine until it closed.

Lawson was hired as vice president of Rocky Mountain Fuel Co. and for 11 years worked for his former enemy. In his new position he worked to improve conditions in the mines and continued to be an advocate for the miners. [18]

Northern Field miners were going through their own years of violence and warfare with the guards at the mine compounds. Conflicts, shootings and even several deaths were reported in all communities during the years of the strike.

Families were not immune from the conflict. In November 1910 school children in Lafayette were reported to be fighting. There was a riot in Louisville in December 1911, and the next day the wives of the miners rioted.

Sixteen men from Lafayette were arrested in June 1911 and sentenced to one year in jail. Their arrests came because they had been congregating on Lafayette streets. Their sentences were commuted after appealing to a higher court and they returned to Lafayette. In the meantime there had been anxiety and hardship on the part of the families.

By December 1911 two men had been killed in Boulder County, riots were reported and shots were fired. After the riots Sheriff M P Capp called for troops to move into the Northern Field.

The settlement of March 1912 with American Fuel Co. operators did not stop the warfare as 2,000 miners were still on strike and strikebreakers were still keeping other mines in the Northern Field in operation. Rocky Mountain Fuel Co. continued to refuse to negotiate.

Foreign strikebreakers in the compound at Simpson Mine, the Rocky Mountain Fuel Co. mine at Lafayette, fired on two union miners. The Lafayette Leader of

Lafayette miners sent to Denver jail

Wives and children visited Lafayette miners jailed in Denver in June 1911. Arrested for "pummeling strike breaker Albert Crabb" were John Bolton, Guy Gordon, Local President Ed Doyle, Frank Koenig and Joseph Carr. Arrested for contempt were Wm Crawford. Walter Gaboury, Andrew Burt, Robert McBernie, Grover Wiseman, K. Gardygis, George Bagdonas, Martin Koenig, John Newcomb, Daniel Williams and Cochines. *Photo from collection of Lafayette Miners Museum*

September 1912 reports that 1,000 shots had been fired. The union miners were said to have thrown rocks through a window in the compound and the non-union miners had then rushed out shooting. Sheriff Capp sent deputies to patrol the streets of Lafayette to try to settle tempers.

With help from the deputies and local law officials, affairs in the towns of the Northern Field settled to skirmishes and complaints against the operators.

In January 1913, when Boulder County Sheriff S. T. Buster took office from M P Capp, the conflict appeared to be more between competing coal operators than miners and guards or scabs.

American Fuel Co. mines were open and operating with nearly a 1,000 union miners. Rocky Mountain Fuel Co., still refusing to negotiate to settle the strike, was using strikebreakers and operating behind barricades.

When Rocky Mountain Fuel Co. reportedly began to cut coal costs for Boulder contracts, American Fuel Co. and the miners worked out a plan to keep the mines open and the miners at work. With negotiating help from John Lawson, and support of United Mine Workers of America, the union miners voluntarily agreed to work for 50 percent of their usual wage. UMWA headquarters made up the difference for the 50 percent wage donation.

American Fuel Co. could compete and keep mines open, miners were working, and Rocky Mountain Fuel Co. was not in a position to control the coal market.

While tempers simmered in the north, conflicts in the Southern Field began to boil.

Ludlow

Southern Coal Field mines were located between Trinidad and Walsenburg and included 38 camps west into the mountains. Bituminous coal was converted to coke and was 60 percent of the state's coal production.

John Osgood was the first president of the operating company, Colorado Fuel Co. A tyrannical operator, he thought mine accidents were caused by carelessness and always the fault of the miners. His mines were closed mines with no union representation allowed, and his workers and their families lived in camps with crude housing.

Even though Osgood's mines and camps were primitive facilities, he had a theory that mine communities should be pleasant places in which to live so that high morale among the workers would lead to high production in the mine. As an experiment Osgood planned a mining community with neat cottages, schools and churches. Osgood's experiment led to the opening of a new mine in 1902 on coal land in a beautiful mountain valley along the Crystal River in Garfield County. It became his retreat and he built the $1 million Redstone Castle.

By 1910 eastern industrialist John D. Rockefeller had acquired control of Colorado Fuel Co. and as an absentee owner had no interest in living or working conditions in the Rocky Mountains. The mine camps had company stores which provided the only source for provisions. Houses were more hovels than homes, and there was no sanitation. Entertainment was one movie a week and no news-

Ludlow after the fires

Chavez Hole after fire on April 19, 1914, at Ludlow Tent Colony in Southern Field.

Photo from collection of Lafayette Miners Museum

papers were allowed. Meetings and even gatherings in the lanes of the camps to discuss wages were forbidden.

John Lawson came to the Southern Field and began talking with miners about the benefits of the union. They were quickly convinced and on September 23, 1913, the general strike was called. Miners and families left the mining camps for even more primitive tent villages, and it was September 29 before the arrival of the 1,000 tents promised by the UMWA. The tents were put on bases of wood floors with pits dug under the floors for storage. Ludlow and several other villages were set up.

Their lives in turmoil, the miners were ready for warfare and began to assemble an arsenal of weapons. Unrest, gunfire, killings were the immediate state of affairs. Trinidad residents asked Colorado Governor Ammons to send troops. On October 11 Ammons sent investigators. Miners had already been killed but when a mine guard was killed, Governor Ammons came south by train to investigate. Armed guards stopped the train.

Shootings continued and October 28 Governor Ammons sent in state troops with the order "Shoot to kill." He also offered a peace plan which was rejected and three guards were killed at Ludlow.

Strikers were ordered to disarm and on November 2 the militia cleared the streets of Trinidad. The hated militia was attacked by strikers. On through the winter the warfare continued with both strikers and militia guards suffering casualties.

The Greek colony at the Ludlow Camp and at other camps in the Southern Field celebrated Easter on April 18. On Monday the strikers and militia began a battle that lasted two days. It was never determined who had fired the first shot.

Families began to evacuate from the Ludlow tents. John Lawson was present to help with the effort. Protection for the escape came when a freight train rolled slowly by so the families could walk to safety. Machine guns fired into the tents, strikers responded, and tents were soon burning. There was never an answer as to who had started the fires, but it was admitted in the later investigation that the National Guard members probably helped to spread the fires. The Guard is known to have helped people escape from the burning village.

Tuesday afternoon, when shootings had stopped, eleven bodies, two women and eleven children, were discovered in the pit beneath one of the burned tents. During eight months of warfare, at least 34 union miners and 26 guards and militiamen had been killed.

It was the death of the children and their protectors - the Ludlow Massacre - that shocked Colorado.

Hecla

Northern Field miners responded to the news of Ludlow with rage and the shootings began anew. Rocky Mountain Fuel Co. was operating the only mines working and the company was an absentee operator with headquarters in Denver. The company defied the union and kept the mines open with foreign labor and guards to surround the compounds. It was easy to fight an absent operator.

Louisville Town Trustees met in a special meeting on April 24, 1914, to take steps to calm the town's residents during this renewed crisis.

A committee of women appeared before the Trustees to ask that saloons be closed. The order was given to close saloons at 6:00 p.m. until further notice. Posters were also posted asking that there be no gathering in groups - either on the streets or at the Depot. And people were not to frequent saloons.

Bullets were flying in the Northern Field by Monday, April 27, a week after the Ludlow Massacre. Most people agreed the first shots were from the Vulcan Mine at Lafayette. Machine gun fire then came from guards at Hecla Mine.

Miners were attending a union meeting and rushed to their homes for weapons at the sound of the gunfire. The fight was on. Louisville miners attacked at the Hecla Mine by approaching the compound through an empty irrigation ditch. When the water gates were opened to fill the ditch, the miners continued the battle standing in water.

Machine gun bullets sprayed homes and buildings in Louisville. Families evacuated when they could or retreated to basements for protection. No miners or Louisville residents were killed or injured by the guards' bullets, but one man in the compound died.

By noon on Tuesday the battle was over. Louisville's mayor, Louis Eberharter, and county commissioners negotiated the truce and Federal troops were ordered to Lafayette and Louisville.

Federal troops arrived by train on May 1, 1914, and set up a tent camp east of Louisville. Peace was restored and the soldiers were accepted into the activities of the towns.

The remaining months of 1914 were quiet but it was not until December 8 that miners convened and declared the strike ended. The soldiers returned to Fort Robinson in Nebraska.

The strike won a wage increase for strikers but Rocky Mountain Fuel Co. continued to deny UMWA recognition as a bargaining agent.

In an attempt to work with the miners, the operator set up company unions which became known as Rockefeller committees. The committees were useful at first and helped with communications between operator and miners. Eventually that communication broke down and mining conditions worsened.

Columbine

The Columbine Mine strike of 1927 was instigated by International Workers of the World. A more militant labor movement, the IWW, called Wobblies, recruited members around the Northern Field coal mines in the 1920's. The goal was to take members and power away from UMWA. The ultimate plan of IWW was to organize the working class to do away with capitalism.

IWW was organized in 1905 and called 150 strikes in industries in America. One of their famous strikes was at the Columbine Mine.

Rocky Mountain Fuel Co. operated the mine at Serene, south of Erie. All other mines in the Northern Field had been shut down but Columbine Mine continued in operation with imported miners. They were housed in Serene, a fenced compound with guards posted around the property.

In November 1927 the union miners in all Northern Field communities were idle and many had been congregating at the mine entrance every day for rallies and to picket the company.

November 21 began in the usual way as a crowd of several hundred (estimates are between 200 and 600) was milling around in front of the guards at the compound gates. The day's confrontation might have been determined ahead because Colorado National Guard and several observers were on hand.

Guards challenged the crowd's rights to picket and demonstrate. A miner asked to have the gates unlocked and insults were exchanged. Words became angry words and then a scuffle. Rocks were thrown at the guards, miners rushed toward the gate, and the shooting began. The miners had no weapons, only the rocks, as they had been instructed to check any weapons at the union hall in Erie. As further proof of the one-sided shooting, no guards were wounded. The six miners who died were John Eastennes of Lafayette, Nick Spandavikis of Erie, Edward Jacques of Louisville, Jerry Davis, George Kovitch and Mike Vidovich. Thirty-six miners were wounded.

Josephine Roche, one of the directors of Rocky Mountain Fuel Co., came from Denver at the first word of the massacre. She buried a machine gun in a mine shaft and promised the welfare of the miners would become the company's concern. She did not have the support of the Board of Directors in her actions but she was able to fulfill her promise.

Contracts for the coal miners had to be upgraded and strikes were called periodically until coal mines stopped operating. But there was never again the violence and death because of the strikes.

United Mine Workers regained union power in Northern Field. In 1955 Union Mine Workers Hall in Louisville was demolished and the Louisville Local was moved to Erie.

The Wobblies' influence in Colorado has disappeared but the national union has not disbanded. IWW representative Gary Cox worked with members of Erie and Lafayette Historical Societies to erect a marker in memory of the union fight and of the six miners who died at Columbine Mine. The memorial plaque was dedicated June 10, 1989, and is located on the south side of Highway 7, about four miles east of Lafayette and 2 1/2 miles east of County Line Rd.

In the 1990's there were still 1,000 members of International Workers of the World. They were working for the original goal.

Columbine Mine miners

Miners from Columbine Mine, all older than 60 years, were photographed about 1937. Left to right: Joe Coss of Erie, Joe Beranek of Lafayette, Ralph Davies, John Quella, Joe Kutcho, George Brown, Arlo Wilson. Front row: Walt Themder, Joe Panek, Jack Scottie Maxwell.

Photo from collection of Lafayette Miners Museum, given to Museum by Louis Gaboury

Northern Colorado Power Company

Louisville-Lafayette Power Plant, located between the two towns, began operation in 1907. The Electric Mine opened the same year to produce coal to power the plant for electricity generation. The mine was worked for only one year but reopened for production from 1914 to 1918. The vein was at 206 feet and produced 74,000 tons of coal. The power plant became the Northern Colorado Power Co. and in 1908 began to generate the 11,000 volts of electricity needed to operate the Interurban train. The passenger train ran between Denver and Boulder by way of Louisville. By 1911 the power company had a transmission line 150 miles long for towns from Boulder to Greeley. Company fortunes began to decline in 1914 because of technical problems within the plant. Also contributing to the decline were depressed economic conditions following the long strike. The final economic blow to the company came Labor Day, September 20, 1920, when a tragic wreck occurred on the line near Globeville. Thirteen people were killed and many were injured. The company was reorganized into Western Light and Power Co. and in 1923 merged with Public Service of Colorado. Valmont Power Plant took over as a major producer of electricity in 1927. The huge plant, which was located north of South Public Road, was dismantled and demolished in 1948.

Chapter 6

The Supporting Cast in the Northern Field History

The coal mines were the story. Without the coal, the railroads to move the coal, and the operators who had the funds to open and run the mines, the Northern Field would not have been part of the history of Boulder and Weld counties.

The miners were the characters of the story, but unions directed the action. Unions and the organizers determined when the miners would respond to the call of the mine whistles signaling work for the day. If the time had come to strike for higher wages, or the issue of that time, the union miners responded. The record says 3,000 of the Northern Field miners were union. Not all miners were, however. The non-union miners were captives of the union movement in the early years of the industry. If a strike was called they stayed off work as well but without the union allotment of funds to keep the family going. If non-union miners chose to continue at work, whether for political reasons or to support a family, they were considered a strike breaker. Up to the 1920's this was a serious charge. From 1930's to the end of the mining era, a coal miner was a union man.

The story of the strike breakers would be a different sort of story. Mine operators hired men to replace the union men on strike. The recruiting for laborers went as far as Europe and many miners were foreigners imported to be able to keep mines open. For the foreigners, even though they were unable to speak English, a job to work in mines in Colorado was an opportunity. They were eager to come to America and families followed as soon as they received their money for passage. If they had to live in a closed compound in company housing, so be it. Life has always been better in America.

Some foreign miners joined the union cause when they understood the language and the strike action. But not always. Some foreigners continued their work in the mines to be able to support their families. They put up with the taunts, epitaphs, the calls of scab. The wives and children did also.

Without the women, unsung, overlooked in history, the towns of the Northern Field would not have been civilized.

Mines were the exclusive environment of men. Women did not even enter mines let alone work there. Two cases have been reported when women were escorted into the mine for a chance to look around. The visits were on days when the mines were not in operation and no men were working. Miners might not call this objection to women in their work environment a superstition, but it was standard practice.

While miners worked in 50 degree temperatures winter and summer, wives of the coal miners put up with summer heat while working in long dresses before fashions changed. Winters they worked in drafty houses. Good thing there were meals to prepare and bread to bake so the coal-burning cook stove could be fired up for heat. That would be in summer too. But some houses had summer kitchens where the heat of a cook stove was out of the living quarters.

Women worked in homes without conveniences and struggled year round with finances. During winter months they budgeted and saved so there would be household money to help the family through summer layoffs and times of strike.

They listened with dread when the mine whistle sounded during the work day to signal accident and disaster. They knew by the sound, each mine whistle being unique, if it was their family's mine.

In those cases when the whistle meant a tragedy and a miner had been killed, the mine owner was not judged to be responsible and was without liability. Surviving familes were given a small settlement, maybe as much as $5,000, but the widows and children were without other resources. The children found what work they could and entered the mines at the earliest opportunity to work with relatives. The widows worked at home, sometimes doing washing and ironing for others, and often were married for a second time to an eligible bachelor.

Women were at the heart of the school system even though the men served on Boards of Education. Women most often were the teachers. Schools were some of the earliest buildings in the area and school districts were the earliest governmental units. Women demanded the opportunity of education for their children.

Church women worked together to better their communities. Women in every Northern Field town have kept church doors open by cooking and serving church dinners to raise funds. Methodist Church women in Louisville organized Christmas treats during hard times. They held fund-raising dinners on election days and their Fall Festivals and well-baby clinics grew into Labor Day celebrations. St. Louis Catholic Church women held dinners and bazaars for decades and raised the money to build their church and school.

Twice Louisville women ventured into men's politics when committees approached Town Trustees. The women asked to have saloons closed, once in 1914 during the strike shootings and once in 1918 in the flu epidemic. And they were.

Mine operators were caught in the strike net. Local companies talked, negotiated and reached a strike settlement as soon as possible. Keeping the mines working was the goal. Enough profit to maintain operations and pay for the investment had to be worked out by the operators and were part of the wage equation.

Mines could be opened without government regulation and the more mines were opened, the slimmer the profit margin for all. Small mine companies were often absorbed by larger and stronger companies if the small mines' coal deposits were extensive. In 1911 Rocky Mountain Fuel Co. became one of the largest mine operators in the Northern Field.

Chapter 7

Rocky Mountain Fuel Co.

Organized in 1890, Rocky Mountain Fuel Co. began operating mines in the Northern Field when it acquired the properties of Northern Coal & Coke Co. in October 1911. Several of the properties were not operating because of the strike which had begun in 1910, and full production was not resumed until 1914 when the strike was settled.

Even when all of the company mines had begun operations the company found it difficult to produce a profit because of competitive pricing from the American Fuel Co. and the Southern Field mines. Rocky Mountain Fuel Co. mines had been damaged and equipment destroyed during the strike. On top of the heavy expenses, bonds had been issued in 1913 to acquire more properties in Colorado. Not only were not all the purchases completed but the bonds were not retired or paid off. Bond interest grew and when the company's debt became too heavy, money was borrowed to pay bond interest. Coal production was high in the 1920's but the company's mines could not turn yield into profit.

In 1927 the president of the coal company died and his daughter, Josephine Roche, inherited his stock and chose to become involved in the leadership of the company. In March 1927 she was able to name three directors to the board and they appointed Merle Vincent as manager of the company. He had a liberal viewpoint and shared with Josephine Roche concern for the miners' welfare but they had not had the power to change the company's policies before the strike at Columbine Mine .

Josephine Roche

Josephine Roche was born in Omaha, Nebraska, in 1886, the daughter of John and Ella Roche. She attended private schools and graduated from Vassar College. She earned her Masters Degree in sociology from Columbia University in New York City.

The Roche family moved to Denver in 1906 and Miss Roche followed after her degree studies. Her first job as a probation officer for juvenile court developed into a concern for children's welfare and child labor laws. She returned to New York for research into the subject for a PhD but in 1912 accepted a job offer in Denver before her degree work was completed. She was Denver's first policewoman as inspector of amusements. She policed dance halls and the red light district until her activities for social betterment came into conflict with the city administration.

John Roche had become president of Rocky Mountain Fuel Co. and was prejudiced against the unions. Since he lived away from the mining operations he had

Early view of Rex #2 Mine, Louisville

The buildings in this early view of Rex #2 were rebuilt when Rocky Mountain Fuel Co. became the owner in 1911. Miners in the photo include Judge Wisson, Lew Wilson, Chris Rosenbaum, James Fullerton, William Machin and Jim McGill.

Photo from collection of Louisville Public Library and Louisville Historical Museum

no knowledge of or contact with miners. Miss Roche had great concern for all laborers and, in spite of her family background and comfortable circumstances, she supported the union movement.

When she heard about the shootings at Columbine Mine she hurried from Denver to the scene. She stepped in to silence the guns and spoke to the miners about her interest in their welfare and her campaign to improve their conditions. The Board of Directors was angered at her interference into the operations and Josephine Roche set about to change the company. She was able to buy out one stockholder and acquired enough stock to become the majority stockholder with 51 percent of the shares.

In March 1928 a new Board of Directors was appointed and Merle Vincent was named president of the company. Josephine Roche was named vice president and she brought John Lawson into the company as a director. Lawson, who had once been the union enemy of Rocky Mountain Fuel Co., had the full support of Josephine Roche and was named a vice president. He served for eleven years in that capacity and was an advocate for the miners. The company began to view mining as a mutual enterprise and should be operated not only for profit but also for the benefit of the miners.

The company invited UMWA to be in the mines and a contract giving recognition to the union was signed in September 1928. Collective bargaining was intro-

Hecla Mine, Louisville
Hecla Mine opened in 1893 and worked two seams until 1920. The mine was 180 feet deep and produced 1,427,137 tons of coal. Hecla Coal Co. opened the mine and was owned by Citizens Coal Co. before being bought by Rocky Mountain Fuel Co. in 1911. Hecla was the only RMF company mine open during the strike years of 1910 to 1914.
Photo from collection of Louisville Public Library and Louisville Historical Museum

duced and wages were raised to $7 a day. Efficiency increased with morale soaring and the company became one of the largest coal companies in Colorado. More importantly, the mines began producing more coal than non-union mines.

Labor problems were solved for the miners but the financial problems would not disappear. During the 1930's the coal industry began to decline as the Great Depression affected Colorado. It was also at this time that the development of natural gas began to reduce the demand for coal. Competitors moved to improve their position in Colorado coal mining and in 1931 Colorado Fuel and Iron Co. in Southern Field dropped wages to $5.22 a day.

Josephine Roche wanted social reforms and tried to win the Democratic party's nomination to run for governor of Colorado in 1934. Ed Johnson won the primary election to run for a second term and Miss Roche gave up politics. She was a supporter of Franklin D. Roosevelt and the New Deal, and she accepted a position in Roosevelt's administration as director of the Public Health Service. Later she was director of the National Youth Administration and was in Washington, D.C., from 1935 to 1937.

It was in 1933 that Congress passed the National Labor Relations Act which gave workers the right to bargain through their unions. Colorado coal miners joined the miners of Rocky Mountain Fuel Co. in union membership and representation.

Paul Peabody was president of Rocky Mountain Fuel Co. from 1934 until his death in 1937. Josephine Roche resigned from her position in Washington and returned to Denver in 1937 to become manager of a company which had fallen into critical financial difficulties. In October 1935 UMWA had made its first loan to the company to meet bond interest payments. Josephine Roche borrowed money from friends to repay debts during the financial crisis.

The union continued to give financial support to the company. Lewmarken Inc. was a trust that was set up by UMWA to help out companies that had been friendly to labor. The company had opened Colorado coal mines to union organization and benefited by promissory notes from UMWA for $450,000. The trust eventually owned 23 percent of the stock in Rocky Mountain Fuel Co.

In spite of the financial assistance from UMWA, in spite of the infusion of cash from friends, the fight for financial stability was lost and Rocky Mountain Fuel Co. declared bankruptcy in 1946. The company was ordered to liquidate assets under federal bankruptcy laws. All mines ceased operation but the liquidation was not completed

Miss Roche continued in control of the defunct company and the remaining assets and moved to Washington. In 1947 she became an assistant to John L. Lewis, president of United Mine Workers of America. She was named director of the Pension Fund in 1948. Josephine Roche became president of Rocky Mountain Fuel Co. in 1950 and its non-liquidated assets. She maintained control until 1971 and died in Washington, D.C., in 1976.[(19)]

Gerald Armstrong became the advocate for minority stockholders in the company and in 1972 conducted a proxy fight. He was able to get the Lewmarken rep-

Acme Mine, Louisville
Acme Mine closed in 1928 and was the last mine to operate within, or under, Louisville's Old Town. Louisville Coal Mining Co. bought the mine soon after it was opened by Mr. Stout in 1890. United Coal Co. and Northern Coal and Coke Co. operated the mine before it was bought in 1911 by Rocky Mountain Fuel Co. Acme Mine produced 1,727,282 tons of coal on two level and was 186 feet deep.
Photo from collection of Louisville Public Library and Louisville Historical Museum

resentative to vote the trust's 240,380 shares to his benefit and the Board of Directors named Armstrong president in 1973. He continued to control the company from the Denver office. By careful management of assets and funds for Rocky Mountain Fuel Co. he began to show a profit for the shareholders, many of whom were coal miners. He promised liquidation and suggested that sale of mineral rights to the surface owners would begin. The annual report of 1979 listed net income of $647,000.[20] The company office in Denver remains open in 2001 to manage assets.

Rocky Mountain Fuel Co. closed mines in 1945 and was the last big company to operate mines in Northern Field. After 1945, and until the last mines closed in 1974 and 1975, mines were independently owned.

Coal mining gradually ceased operations. Lafayette's mines were gone by 1942 when the Black Diamond was closed. Marshall's huge district closed down in 1958 when Cuba Mine stopped working. Louisville had a working mine until 1955 when the HiWay Mine closed. Superior's last mine was Gorham which closed in 1955.

Other mines around Boulder County were closed earlier for a variety of reasons. The big Industrial Mine at Superior was closed by trustees of Rocky Mountain Fuel Co. when the grade of coal was determined to be unmarketable. Mine machinery was sold for $50,000.

In February 1946 fourteen Boulder County mines closed putting 427 miners out of work.

Four mines, Centennial, Monarch, Crown and Pluto were closed in April 1947 by the federal government for safety reasons. A strike at HiWay Mine in July 1947 brought a pay raise to $13.25 a day and an eight-hour work day. After a fire that destroyed the tipple in 1954 the mine did not reopen. The Baum Mine at Frederick closed in 1958 when all available coal had been mined.

Erie and Dacono were the last mining communities in Northern Field. The Lincoln Mine at Dacono closed in 1974. Erie's Eagle Mine ceased operation in 1975. Both mines had a workforce coming from throughout the Northern Field.

The mines in Boulder and Weld counties had a total estimated coal production of 100 million tons from 1866 to 1975.

Rex #2 Mine, Louisville

The Rex Mine operated in two seams and both opened in 1898. Rex #1 was worked until 1917 and produced 1,666,247 tons of coal at 148 feet deep. Rex #2 closed in 1915 after production 745,652 tons of coal. Northern Coal and Coke Co. opened the mines and sold the properties to Rocky Mountain Fuel Co. in 1911.

Photo from collection of Louisville Public Library and Louisville Historical Museum

State Mine, Erie
THE SHIFT PHOTO AT STATE MINE WAS TAKEN ON APRIL 25, 1927.
Photo from collection of Lafayette Miners Museum

Chapter 8

The Coal Not Seen

Mines are closed, tipples are gone, all surface indications of the former industry have been scoured away. The underground world of the mines in Boulder and Weld counties has not disappeared, however. An estimated 216 million tons of the coal deposit remains. Can it ever be harvested? The lignite coal has a favorable sulphur content and would be useful for industry by turning it into a synthetic gas through liquification or gasification.

Most of the mines in Boulder and Weld counties are too deep underground and costs for reopening would be prohibitive. New rules for safety make obsolete the mining methods formerly used here. Since the deep mines won't be opened for traditional mining that leaves surface or strip mining as a way to reach the remaining deposits. Especially around Marshall veins of coal can be seen rising to the surface.

Surface mining would be as cost-prohibitive as underground mining. The costs for reclaiming the surface to modern EPA standards make that option unfeasible in present-day America, in the unlikely event that environmental concerns would permit any form of surface mining in the first place.

However, America needs fuel and resources are not unlimited. Economics will control any future use of the remaining coal deposits. A bill was introduced in the Congress of 2001 to spend money over the next decade for clean coal research and to give tax breaks to utillty companies and industry to use the new technology. Turning coal into gas is a technology of the future. Retrieving the coal is a developing technology and mining machines controlled by computers on the surface are already at work in Kentucky coal mines.

There are coal reserves available in Northern Field and the coal is a quality useful for industry. Billions of dollars are being spent to develop the technologies to retrieve coal and make use of it. Northern Field may still have a product to offer.

Subsidence

In the coal mines of the Northern Field millions of tons of the coal were removed. In the room and pillar method of mining that was used, pillars of coal or heavy timbers supported the roofs of the rooms where coal was mined. As the coal was removed and the room abandoned, the pillars were either pulled down or gradually mined away to allow the roof and earth above to collapse into the room. With the pillars taken down the subsidence process would begin.

As earth above a closed mine room collapsed into the void, the caving moved upward. When the caving reached the surface, subsidence had occurred. In the deep mines, most often found in Northern Field, caving seldom reached to the surface. In the mines with the pillars removed, the subsidence occurred rapidly and soon after the mines were closed.

In some rooms where the pillars and timbers were not pulled or removed, voids would be left. Roofs and rock falls could have fallen into these voids but there is always the potential for subsidence. It is the potential for subsidence that has to be considered carefully by cities, developers and home owners today. Of the 87,000 acres of inactive mines in Boulder and Weld counties, it is reported that there could be subsidence potential on 20,000 acres.[21]

Historically, subsidence was often unnoticed and not an issue. Northern Field was a rural area and many mines were opened in areas of empty fields. Evidence of subsidence would be seen as swales and depressions on the farmland and these would have been evened out in the process of farming. Damage to small buildings was easily repaired and roadways could be regraded.

The Town of Louisville experienced subsidence in 1902 that was so severe that a large two-story brick building at the corner of Main Street and Pine Street was condemned and had to be demolished. Photographs show the damage as a two-foot drop along foundations in a two to three block area. The Town Trustees considered billing the mine operator for damages to streets and properties. The Acme Mine was worked beneath the Old Town of Louisville on two levels at depths of between 150 to 200 feet. Since its closing in 1928 there have been no reports of further subsidence. Other towns in the Northern Field have had subsidence but most episodes are part of the local history and there is no documentation.

Not until the small towns became cities with new residents and new houses, developments of the past 25 years, did subsidence potential become a concern for city Planning Departments and home owners.

There are a number of factors that have to be considered today when looking at subsidence potential.

Dates of operation:	Establish the dates of operation of the mine and length of time mine has been abandoned.
Depth of Mine:	Deep mines in Northern Field were up to 300 feet deep. Thickness of the vein of coal and the level of cover is also figured.
Roof Stone:	In some mines the roof stone is a layer of iron stone that can be from 12 feet to 80 feet in thickness. A roof in a room may collapse but the upward caving is stopped by the layer of iron stone. The room may be a void under the iron stone.
Floor of Mine:	Floors of sandstone would support pillars and timbers when these were allowed to stand. Soft shale or clay would allow the pillar to sink with eventual caving and subsidence.

Pillars: Presence or absence of standing pillars.

Rock faults: Location of rock faults.

Water in abandoned mines is also a factor in subsidence potential. There is a natural underground water flow through some, if not most, coal veins in Northern Field. The underground water finds its own channel through soils and rock faults and into the mines. The water flow through some soils will even collect more in different surface areas. The water that flowed into the coal mines had to be constantly pumped out in order for work to continue in the veins to remove the coal. When the mine was closed the pumps were turned off and the mines filled with the ground water.

J. E. Turney says in his booklet, Subsidence In Inactive Coal Mines, "Ground water in the mine provides a buoyant force that helps support the ceiling. Also, pillars retain their strength because the lack of oxygen in the water-filled mine prevents the chemical breakdown of the coal. Therefore, water level changes in the mine increase the chance of pillar failure. Changes can contribute to subsidence."

Included in the mine's water factor would be whether the mine water was used as well water. Louisville's water from the deep Acme Mine was not used because the water was too brackish, or salty, for consumption or for watering gardens or lawns. Lafayette has had more wells opened for surface uses because the water-filled mines were not quite so deep and therefore not so salty.

Frederick used the abundant underground water flow for the town's water supply and some mine water was tapped. Farmers in both counties drilled wells into the underground water for field and stock irrigation but there has always been enough water flow to replace any supplies used.

Up to modern times underground water has continued to flow into the mines and water use has not created voids that would lead to subsidence. However, Turney's report on underground water in abandoned mines serves as a warning that water levels in those mines should not change. Mine water should not be tapped in the future for surface uses.

Factors not involved with abandoned coal mines can imitate subsidence damage. Colorado's soil can cause damage to construction that will look very much like the damage caused by subsidence. Rocks and the clay soils of the state will change volume as the moisture content changes. The swelling and shrinking of clay soils can cause heaving in concrete floors and cracks in drywall and brick exteriors. Poor construction practices and even low quality concrete can duplicate subsidence damage and should not be confused with subsidence over abandoned mines.

Laws were passed by Colorado Legislature in 1973 to address the problems of subsidence in areas of abandoned mines.

House Bill 1041 says mine subsidence is a geological hazard and has to be considered for safe development of land.

Senate Bill 35 says all such hazards require geological reports before a subdivision can be considered.

Before developments can be considered in cities and towns of the Northern Field, engineers must prepare the geological report for the developer. Local governments have additional requirements for development and subdivision plans.

The geological reports must consider: If an abandoned mine exists under the site; void spaces have to be determined by drilling; surface evidence of subsidence requires drilling; subsidence effects on proposed development has to be determined; safe building areas have to be determined; unsafe building areas have to be determined and left as open space.

Maps are available in Planning Departments of local governments which show classifications of hazards for the undermined areas of the coal field. Classifications are severe, moderate and low.

Lafayette and Louisville Planning Departments have maps which show all locations of all mines, all rooms, all pillars, all rock faults, all unmined spaces underground in the city and planning area. Subsidence potential is known and if there is a high risk for subsidence then development planning is adjusted for open space and buildings are not allowed.

The State of Colorado has requirements for building over abandoned mines. Local governments have requirements for development over abandoned mines. But home owners must become involved in the process of evaluation for subsidence. City maps should be studied and Planning Departments are helpful. An engineer's report should be ordered before home construction begins and the type of construction of homes and buildings should all be taken into consideration in determining subsidence potential.

The State of Colorado has set up the Subsidence Protection Program for homes built before 1989. Homeowners who have structure damage from subsidence from abandoned mines may be eligible for funds for home repairs. Up to $50,000 could be available for repairs for those properties deemed eligible. A state trust program administers the fund and processes requests. A fee is charged to enter the program and requirements have to be met to determine eligibility.

For more information about the program the telephone number is 1-800-44MINES. The Department of Natural Resources oversees mining in the state, reclamation and subsidence issues.

Publicity in local and area wide newspapers announced the Subsidence Protection Program. As of December 2000 the program had processed 13 claims, all from Colorado Springs.

The mines of Boulder and Weld counties produced over 100 million tons of coal. Production figures in all of Colorado coal mines through 1960 were 500 million tons of coal.

The huge area of coal deposits that is Northern Field created communities to support the mining industry. Thousands of men were employed and many died working in the mines.

Since Northern Field mined only about one fourth of all coal produced in the State of Colorado the colorful history of the northern districts is overshadowed. The hard mining of all metals, especially the silver and gold, has always been perceived as the history of mining in Colorado.

This book presents a short history of coal mining in northern Colorado. Northern Field is remembered.

Hiway Mine, east of Highway 287, Lafayette
Work was begun to open Hiway Mine in June 1930. At a depth of 385 feet, it was the deepest mine in Boulder County. Veins ran four feet to eight feet high. Hiway worked from 1930 to 1954 and produced 2, 334,000 tons. It was locally owned by Louisville-Lafayette Coal Co. and then by Hiway Coal Co. The mine was wired for electricity and a fan was used to improve air conditions. Mechanical loaders were also used. A fire in March 1954 began in the tipple and was thought to have been caused by faulty wiring. The tipple was destroyed and damage estimates were $60,000. The mine closed in 1954.
Photo from collection of Lafayette Miners Museum

Sunnyside Mine, south of Louisville
Sunnyside Mine was opened in 1900 south of Louisville by Big Six Coal Co. Vesuvius Coal Co. bought the mine in 1904. The shaft was sunk to a depth of 324 feet and produced 299,661 tons of coal. It closed in 1921.
Photo from collection of Lafayette Miners Museum

Centennial Mine, Louisville *(Opposite page)*
Centennial Mine, the second and New Centennial, opened 1936 at depth of 285 feet. The mine operated until 1952 and produced 1,834,000 tons. Northern Coal and Coke Co. opened the mine and a local company, Boulder Valley Coal Co., acquired the property in 1948. Centennial was the first mine to replace mules with mechanized equipment and 70 percent of the coal was handled mechanically. The old Centennial Mine worked from 1905 to 1936 at a depth of 280 feet. The mines were located along Highway 42, east of Louisville.
Photo from collection of Lafayette Miners Museum

Black Diamond Mine #2, Lafayette
The Black Diamond Mine opened in 1931 and worked until 1955. Mine production was 759,238 tons of coal. The shift photo was taken in August 1942.
Photo from collection of Lafayette Miners Museum

Appendix

Boulder County Sheriff S. T. Buster gave a deposition of his official part of the events that took place the night of April 27, 1914, in and around the Hecla Mine and in the Town of Louisville.

That I now am and since the 14th day of January, 1913, have been the Sheriff of Boulder County.That I have lived in the County of Boulder since 1879 and during nearly all that time have been engaged in the business of farming; residing at the time of my election as Sheriff on a farm one and a half mile northwest of Hygiene, Colorado.

That the United Mine Workers declared a strike on April 1910 and that the strike was in full force at the time I took my office as Sheriff.

On the 27th day of April, 1914, I had in my office and in the strike field the following under-sheriffs and deputies:

Robert Euler, a resident of Boulder County for 42 years out of his 44 years of life.

Gene Shreeve, who I understand was born in the State of Colorado and has lived in Boulder County for the last 14 years.

Frank Birch, who was born and raised at Hygiene.

B. O. Smith, who has resided in Boulder County for the last 8 years.

Al McDonald, the Town Marshall of Louisville, and who was a Union man, belonging to the Local at Louisville.

Tony Rappa, who has lived at Louisville for a number of years and is a Union man belonging to the Local at Louisville.

Steve Gorce, of Marshall, also a Union man and official who lived at Marshall.

Ivan Creese, born and raised in Boulder County.

Fred Duncan of Boulder, residing there for years.

All these deputies who received any compensation whatsoever were paid by Boulder County.

In dealing with strike matters in Boulder County it has been my policy to take night side, but to do the best within my ability to preserve law and order.

Up to the 27th day of April, 1914, I was frequently assured by the local leaders of the United Mine Workers, as well as many of the members, that the Sheriff's Office and the deputies had always treated the Union men fairly. I saw to it that there was no law to their interest or in their behalf but what was enforced in Boulder County.

When I went into office I found that both sides were carrying arms, when going to the depots or railroad crossings. I immediately started to eliminate this feature as soon as possible, assuring both sides of a square deal, and had practically eliminated that practice from both sides at the time of the trouble at the Hecla Mine.

In conversations, whenever the opportunity presented itself, I assured the Representative of the Union that I wasn't elected for the United Mine Workers, or for the coal operators, but that I was elected for the people, and was standing by them, that I had no axe to grind and their fight was not my fight. I was simply there in my official capacity as an officer to preserve peace and order.

After the trouble at Ludlow and the attitude the Rocky Mountain News took in the matter, and the officials of the United Mine Workers, I realized that things were in a strained condition in Boulder County. All it needed was a fire cracker to go off at the wrong place to start trouble.

From that time on I stayed night and day, in fact practically lived in the coal camps, watching every move.

On Thursday, the 23rd of April, 1914, while at Louisville I had a call from my office telling me to come to Boulder immediately. They didn't seem to be disposed to tell me over the 'phone for what purpose I was wanted. I went in and learned that the strikers had a scheme on to shoot by the Gorham Mine that evening. I immediately 'phoned to Fred Dungan, who had been one of my deputies through the Winter; Under-Sheriff Euler; Gene Shreeve and Ivan Creese, who started from Boulder to Marshall. On arriving at Marshall I stopped at Doctor Walker's place, and while talking with Doctor Walker I saw three women coming down the street carrying some packages. I stepped out and stopped them and asked where they were going and what was the trouble. They said that the Gorham "scabs" were going to shoot up the town and they were getting out of town. I asked them where their men were and they said they didn't know. They could give me no satisfaction regarding where the men were but said that the "scabs" were going to shoot the town up and they were getting out. While talking to them a Frenchman came up and heard me talking to these women. He gave me a punch and calling me to one side and said "Come with me and I'll show you where the men are." We immediately went up into French-town and stopping at one of the houses he gave a whistle. It was answered by another whistle. We went in and found the house full of men armed with rifles. I asked what they meant by that and they said "the scabs are going to shoot us up and we are going to protect ourselves and families and not allow anything to happen like happened at Ludlow." I told them I was elected for that purpose and if there was any fighting going on I was going to do it and didn't propose to have them do anything, but if I found it necessary I would deputize every able bodied man in the town to keep peace and order; assuring them that they were going to get protection and a square deal. I sent them home. That was about 11:00 to 11:30 in the night. I went from there to Superior and Louisville and came back about 1:30 to Marshall and then went on into Boulder arriving there about 2:15.

I had hardly arrived in Boulder when I was notified that there were twelve shots fired into Gorham Mine from French-town; This information came from a good Union man there. He said that the shooting was done by parties who lived close thereby. I tried to catch the parties who did the shooting but was unable to do so.

On Friday evening I went back to Marshall. Realizing the condition of affairs I talked with several of the Union men and found that one Steve Gorce had the confidence of all the Union people. I immediately gave him a commission as a Deputy Sheriff, feeling that in doing this it would eliminate and keep trouble down if anything was likely to start. After giving him the commission I found afterwards that they all seemed to be suspicious of him; thinking he was standing in with the other side making them lose confidence in him.

I stayed out and worked all Friday night between Louisville, Marshall and Superior. Saturday night I went out again on the same mission working Louisville, Marshall and Superior. About 12:15 Saturday Night I was called to the 'phone by Jack Criss, Deputy Sheriff, informing me that there were about fifteen to twenty-five men missing in Louisville and he was afraid of trouble starting at the Gorham Mine. I immediately started for Superior. I stayed out that night until 5:00 in the morning. Nothing happened, but later on I found that there were in the neighborhood of forty to fifty men who had gathered for the purpose of shooting up the Hecla Mine. One of the Union men found it out and told them if they did any shooting he was going to turn everyone of them in. This scared them out for the time being.

During all this time I was in daily communication with George Carlson, District Attorney, at Denver, and Governor Ammons; who cautioned me to watch every move and be careful, that conditions were such that it required a whole lot of good judgment and good work if we kept trouble down.

On Monday evening, the 27th of April, I received a telephone message from George Carlson, informing me that the operators, through their attorney, and the United Mine Workers, through their Attorney, and himself had reached an agreement whereby the Sheriff's office was to take charge of machine guns and searchlights at various mines belonging to Rocky Mountain Fuel Co. in Boulder County.

I called Joe Potestio, of the Louisville Local of the United Mine Workers, and informed him of the agreement which had been reached between the Attorney for the United Mine Workers and the Attorney for the Operators, and Assistant District Attorney Carlson. Also called up Mayor Faull of Lafayette and informed him of the agreement which had been reached in regard to the Sheriff's office taking charge of the machine guns and searchlights of the Rocky Mountain Fuel Company and that a Deputy was to be placed in charge of each mine, which seemed to please him and he expressed himself as being very well satisfied. I told him I had sent a Deputy over that evening to take charge of the Mitchell Mine and as soon as I could arrange it would get a Deputy and send him to the Simpson Mine and one to the Standard and that I was going to the Hecla Mine and Ivan Creese to the Industrial Mine at Superior, and Gene Shreeve and Fred Dungan to the Gorham Mine at Marshall.

I left Boulder on the 6:30 car for Louisville. Arriving at the Union Hall about 9:00 I found 400 men gathered there to hear what I had to say in regard to the agreement which had been reached between the United Mine Workers' attorney and attorney for the Coal Operators. Part of them seemed to agree to it and think it was a good thing, but some of them had some complaint to offer; while the secretary of the Local, Joe Potestio, made a strong anarchistic talk against me taking charge of these guns, saying that with me being alone, and the deputies at the other mines also, they might tie us and gag us and shoot up the town and kill the Union people. He made a very exciting talk and seemed to arouse the people very much.

One assertion he made was that if Sheriff Buster would say the word they would go down to Hecla and that the (sic.) machine gun and they all jumped up and said "yes, yes." I called him down several times and said it was up to the conservative people to keep down the radicals as it needed but little to start trouble. The conditions were discussed pro and con up to about 10:00 o'clock. During the meeting they waved the Rockefeller flag a great many times. About 10:00 o'clock I asked Pat Powers, President of the Local, to adjourn the meeting, as it was getting late, realizing the fact that the talk Potestio had made had excited the men and I wanted to get ahold of Potestio and talk with him. Immediately after the meeting adjourned I talked with him and told him I didn't like his talk and that it was dangerous on account of conditions being such as they were at that time. He agreed with me and I told him I would like to see him in his office and also told Pat Powers I would like to see him, he being President of the Local. We then went to the office of the Board Members, some six or seven of us. We hadn't any more than gotten into the office and started talking when the 'phone rang. I was called to the 'phone by Mr. Belk, telling me that they were shooting up the Vulcan Mine, which is situated south and west of Lafayette and straight east of Louisville. I turned to the officers in the office and told them what was going on. They all turned white from fright. I said to Potestio "I want you to go with me" and to Pat Powers "You go down there." Potestio said he wouldn't go and the others said they wouldn't go, but Pat Powers said "I will go with you."

We immediately started out and as I got outside I could hear the shots at the Vulcan and the whistle started to blow for help. One of the men in the crowd standing around the Union Hall started for the Town Hall to ring the fire bell. Powers hollowed and told him not to ring that. Jack Criss, my Deputy, was standing at the door as we went out. I told him what was happening and said "Come on Jack let us go to the Hecla." Powers was with us following behind. When we got about half way from the Union Hall to the Depot it occurred to me to have Powers call up Lawson and Doyle. I stopped and told him to go back and 'phone Lawson and Doyle and tell them they were shooting up the Vulcan Mine, and they were then starting to shoot in Louisville towards Hecla, and if any one was killed I would hold them responsible, as well as for any act of violence. He said he would go back and call them. He immediately started. Criss and I started for the Hecla Mine which is about a mile from the Union Hall. When we got to the depot we started north up the C&S RR track and went only a half a block when some person stepped out from behind one of the saloons and emptied his rifle at the Hecla Mine. By that time shooting was going on all over town and the men made for their homes, got their rifles and ran out into town and pointing their guns toward Hecla turned them loose. We had only gotten half way when it was a continuous roar all over town from the shooting.

The roads being muddy when we got to Jake Williams' corner we cut across the prairie north and east towards Al McDonald's place to get him. My hopes were to get up there and head them off and to keep the men on the inside of the stockade from shooting back. I got Al McDonald and we started on for the Hecla.

When we got to within 50 feet of the corner on which is located the arc light of the Hecla property, a party of Union men came in from behind, about 200 yards south of us, and commenced to shoot at us, and about that time they turned loose from the inside and we were between the two fires. We turned back into a house on the corner to get out of the range of the bullets, for by that time the Union forces had gathered along the west side of the railroad track, running north, and south behind the houses down in the north end of the ball grounds. There was a continual roar of shots. We were in this house perhaps ten minutes before the machine gun did any shooting. We were all of twenty minutes coming from Union Hall up to this house. It was about 10:20 when the shooting commenced and there was a continuous roar from then on, up to about 3:15 or 3:30 in the morning.

When I dropped back into this house I found a telephone there and called up the stockade, telling them I didn't want them to shoot into the town and that I didn't want any women and children killed and to cut it out as I would stand for no shooting, and that I would get in there as soon as there was a let up, but all this time, up to 3:30 in the morning there was a continual firing from the town into Hecla Heights.

About 4:00 o'clock things being quieted down and there being no shooting I called up the guard at the mine and told him I was coming in. Mr. Criss and myself started for Hecla Heights. We had hardly gotten to the guard house at Hecla Heights when the shooting started again, more severe than before. It was about 4:00 in the morning when the second shooting was started. I told the men at the Heights that I didn't want them to shoot out into town and they agreed that they wouldn't. I told them that there was no use shooting at those people; that they had ammunition to burn up and would not be satisfied until they had burned it. The only thing to do was to take care of their families and get down in the basement if they had any. All the machine gun was used for was to scare the men back in the ditch as I didn't want any one killed, but if they started to make a rush to shoot to kill. As long as the strikers stayed in the ditch I felt, though they were trying to kill us, they were accomplishing nothing and only showing us where they were located.

While I was in the Hecla guard house, about 8:00 o'clock while 'phoning a bullet came crashing through from a northwesterly direction and lodged in a little cupboard on the south side, just below my knee, just missing my leg.

From 4:00 A.M. until 12:30 I was back and forth between the Rex No. 1 Mine and dump and Hecla guard house.

On arriving at Hecla I found that the men only had about 19 rifles and shot guns and they were not all in use as some of the men had discarded them and gone to the cellars. They had about 15 rounds of ammunition to the man.

The shooting stopped between 12:10 and 12:30.

About 10:00 o'clock in the morning I was informed by one of the men at the Hecla Mine that there was one of the workmen shot in the jaw and one shot in the hand and wanted a Doctor. I called up Dr. Wolfer and told him I wanted him to come over. He said he couldn't get over there that it would be taking his life in his hands for if he started they would kill him. All that we could do under these conditions was to wait and hope for the best.

When the shooting stopped, about 12:10, Mr. Saulters, one of the guards, and myself started up into the boarding house to see how badly these parties were injured. After looking over the situation at the boarding house, and talking with the parties injured we then went over to the Bulgarian quarters and found a Bulgarian lying on the floor on a mattress with some old quilts for a pillow, practically dead, with a bullet hole in his head, entering from the right side on the back and lodging on the left side. These were all that we found wounded. One woman, however, in one of the little houses had a lock of hair shot off.

The condition of Hecla was one of the saddest sights I have ever seen: The boarding house was practically all shot to pieces and had blood all over; the women and children were in the basement hysterical and crying; and down at the Bulgarian headquarters the poor Bulgarian lying shot and dying. The people were stampeding, wanting to know if I couldn't get them out.

About that time Harold T. Martin, Deputy Dist. Attorney, Jack Clark, Sam Greenwood, County Commissioners and Mr. Tedrow, County Attorney, in company with Louis Eberhart, Mayor of Louisville, arrived on the scene.

I took them over the battle ground and showed them the sights: First to the Bulgarian headquarters, showing them Pete Steinoff shot through the head lying there dying; then to the boarding house and into the basement, showing them one workman with his jaw practically shot off, another with a shot through his hand lying there on some old quilts; the basement was filled with women and children. I then took them up into the washroom in the boarding house; then through the boarding house, showing them the conditions. They said it certainly "got their goat."

We went from there to Louisville.

The transcription of Sheriff Buster's deposition was among the papers in the files of Rocky Mountain Fuel Co. Historical papers and files of the company have been given to Western History Collection of Denver Public Library and the original transcription is now in the Western History Collection.

Acknowledgements

Thank you to all the people who have given help and information so willingly, not only for this book but also for all past written endeavors.

The Louisville Historical Museum, Director Carol Gleeson and Todd Gleeson, and the Historical Commission; Louisville Public Library; Lafayette Miners Museum, Director Sloane Schwindt and the keepers of the history of Lafayette, Jim and Beth Hutchison. All helped and were generous in giving time and photographs.

Government officials were helpful in (short) coversations and provided information so we would all know the ramifications of the mining:

City government departments: Hallie Sawyer of Erie; Nancy Guynn of Lafayette's Planning Department and Peter Johnson of Public Works; Tom Phare of Louisville Public Works.

Jim Cappa and Jim McArdle of Colorado Geological Survey, Division of Active and Inactive Mines (Coal Section)

Loretta Pineda of Colorado Division of Minerals and Geology.

Gerald Armstrong, President of Rocky Mountain Fuel Co. *(Not a government official)*

Louisville resident Frank Domenico helped with maps and lists of mines. The reports and information he gathered have been given to Louisville Historical Museum.

Many people all through Northern Field listened, answered questions and helped in gathering details: Helen Susuras of Dacono, Bill Flynn of Dacono, Ron Zaruba of Erie, the Schreiters, Donna and Dave of Louisville and Becky Schreiter Campbell of Frederick, Violet Bonnelli Gardner of Louisville, who looked up information and photos, Eleanor McGinn of Lafayette, whose biography of Josephine Roche will be published by Colorado History Museum, and Joanna Sampson of Marshall who once again offered helpful information and history.

The late Andy Deborski of Louisville shared his experiences and knowledge about coal mines not only with me in the years when I first was gathering information but also with others in local schools and the museum. Andy always had time to talk about the coal mines. He died November 22, 2000.

Mike Campbell of Longmont, with his design skills, has arranged the copy, charts and photographs into a pleasing text. Because of him it is better, and it was much easier for me.

Cynthia Campbell of Longmont edited and controlled the commas.

As always, thanks to my patient husband and family.

Notes

1 Jay Gould was a share holder in Colorado Central Railroad in 1872. In 1879 Jay Gould gained control of Missouri Pacific Railroad.

2 Sources refer to Denver, Marshall and Boulder Railroad. Colorado Railroad Museum Map has no reference to D. M. & B RR. Rick Steel's web site for Colorado Southern and Union Pacific Railroad History reports D. M. & B. associated with Colorado Central and Colorado Southern Railroads.

3 Sources refer to Union Pacific, Denver & Gulf Railroad or to Denver, Gulf & Pacific Railroad. Map of Colorado Railroad Museum has no reference to D.G. & P or U. P. D. & G. Railroads.

4 "History of Erie United Methodist Church," Dave Olson. "Erie: Yesterday and Today," Social Studies and History classes of Erie High School.

5 Charles Spencer was a partner in the company and Simpson Mine was also known as the Spencer Mine.

6 Lafayette Leader, Nov. 29, 1912. "Mary Miller: Boulder County Pioneer."

7 Lafayette-Louisville Electric Co. became Northern Colorado Power Co.

8 "Superior: A Folk History." Superior Community Involvement Program.

9 "The Great Coalfield War." George McGovern and Leonard Guttridge. Houghton-Mifflin, 1972.

10 "Erie: Yesterday & Today."

11 Denver Post articles: Jan. 20 to 23, 1936. "After damp" is a toxic gas remaining after an explosion.

12 Forming UMWA 1890 - Knights of Labor. UMWA internet web site: A Brief History of UMWA.

13 "Unionization in Boulder and Weld Counties to 1890." Dennis Grogan, Colorado Magazine, Fall, 1967.

14 Rocky Mountain News article: reprinted in Lafayette Leader Jan. 14, 1901.

15 "The Great Coalfield War."

16 "Mother Jones: Miner's Angel." Dale Fetherling. Southern Illinois University Press, 1974.

17 Lafayette Leader. March 8, 1912.

18 "Out of the Depths." Barron Beshoar. Golden Bell Press, 1958.

19 "Josephine Roche: Social Worker and Mine Operator." Marjorie Hornbein, Colorado Magazine, March 1976.

20 Rocky Mountain News, April 11, 1973. Morton Margolin

21 "Their Silent Profile: Active and Inactive Coal and Metal Mines of Colorado." Reclamation Division, Colorado Department of Natural Resources.

Sources

Books

McGovern, George and Leonard Guttridge. "The Great Coalfield War." University Press of Colorado, 1996. Houghton Mifflin, 1972.

Beshoar, Barron. "Out of the Depths: The Story of John R. Lawson, A Labor Leader." Golden Bell Press, Denver. 1958.

Fetherling, Dale S. "Mother Jones: The Miners Angel." Southern Illinois University Press, Carbondale and Edwardsville, 1974.

Hafen, LeRoy R. "Colorado and Its People." Lewis Historical Publishing Co., Inc. New York, 1948. "The Labor Movement," Oliver Dickerson.

Sinclair, Upton. "The Coal War. Colorado." Colorado Associated University Press, Boulder, 1976.

Ayer, Eleanor. "Green Light on the Tipple." Platte Press, Frederick, Colorado, 1978.

"Superior: A Folk History." Superior Community Involvement Program.

"Their Silent Profile: Active and Inactive Coal and Metal Mines of Colorado." Mined Land Reclamation Division, Colorado Department of Natural Resources. 1982.

Pamphlets

"Erie: Yesterday & Today." Erie High School Social Studies and History classes.

Olson, Dave. "History of Erie United Methodist Church." 1998.

Peck, Maynard. "Some Aspects of the Coal Industry in Boulder County, Colorado." Thesis for Department of Economics, University of Colorado. 1947.

Turney, J. E. "Subsidence Above Inactive Coal Mines." Special Publication 26, Colorado Geological Survey and Colorado Mined Land Reclamation Division. 1985.

Sources

Reports

Coal: 1970. Coal Mine Inspection Division, Department of Natural Resources, State of Colorado.

Annual Reports of Coal Mine Production, State Coal Mine Inspector, Coal Mine Inspection Division, Department of Natural Resources, State of Colorado.
Dames & Moore Engineering Report: Boulder County Subsidence Investigation, 1986.

Meyer, Erwin. "A Case Study of Industrial Conflict." The Survey, issue of February 15, 1928. Western History Collection, Norlin Library, University of Colorado.

Map of Colorado Railroads. Colorado Railroad Museum, Golden.

Newspapers

Colorado Prospector, October 1975. Gwen Goldsberry, editor.

Louisville Times, Colorado. Selected issues.

Boulder Daily Camera, Colorado. Selected issues.

Lafayette Leader, Colorado. Selected issues.

Denver Post, Colorado. Selected issues.

Boulder News and Courier, Colorado. Selected issues on file Western History Collection, Norlin Library, University of Colorado.

Series of articles, dated 1940, in Louisville Times written by Bill Marshbanks on Louisville mines.

Hecla Mine after closing in 1920.
Rocky Mountain Fuel Company photo from Louisville Public Library and Louisville Historical Musem.